FINGERPONDS: SEASONAL INTEGRATED AQUACULTURE IN EAST AFRICAN FRESHWATER WETLANDS

Exploring their potential for wise use strategies

Promoter	Prof. dr. P. Denny Professor of Aquatic Ecology UNESCO-IHE/Wageningen University, The Netherlands
Co-promoter	Dr. A.A. van Dam Senior Lecturer in Ecological and Environmental Modelling UNESCO-IHE, The Netherlands
Awarding Committee	Prof. dr. F. Kansiime Makerere University, Uganda Prof. dr. B. Moss University of Liverpool, United Kingdom Prof. dr. ir. A.J. van der Zijpp Wageningen University, The Netherlands Prof. dr. P. van der Zaag UNESCO-IHE, The Netherlands

Fingerponds: seasonal integrated aquaculture in East African freshwater wetlands
Exploring their potential for wise use strategies

DISSERTATION

Submitted in fulfilment of the requirements of
the Academic Board of Wageningen University and
the Academic Board of the UNESCO-IHE Institute for Water Education
for the Degree of DOCTOR
to be defended in public
on Wednesday, 20 December 2006 at 13:30 hours
in Delft, The Netherlands

by

JULIUS KIPKEMBOI
born in Uasin Gishu, Kenya

CRC Press
Taylor & Francis Group
Boca Raton London New York

CRC Press is an imprint of the
Taylor & Francis Group, an **informa** business
A BALKEMA BOOK

First issued in hardback 2017

CRC Press/Balkema is an imprint of the Taylor & Francis Group, an informa business

© 2006, Julius Kipkemboi

Published by:
Taylor & Francis/Balkema
PO Box 447, 2300 AK Leiden, The Netherlands
e-mail: Pub.NL@tandf.co.uk
www.balkema.nl, www.taylorandfrancis.co.uk, www.crcpress.com

ISBN10 0-415-41696-5 (Taylor & Francis Group)
ISBN13 978-0-415-41696-2 (Taylor & Francis Group)
ISBN 90-8504-497-9 (Wageningen University)

ISBN 13: 978-1-138-42434-0 (hbk)
ISBN 13: 978-0-415-41696-2 (pbk)

Table of Contents

Abstract

Natural wetlands around Lake Victoria are threatened by unsustainable exploitation. Poverty and unreliable terrestrial production have led to increased dependence on natural wetlands for livelihoods. The main objectives of this study were to (1) explore the potential of enhancing food production from natural wetlands through a simple and appropriate technology; and (2) investigate the integration of this technology into sustainable floodplain and littoral wetlands farming systems.

Fishponds were dug in the wetlands and the excavated soils were used to create raised bed gardens beside the ponds. The ponds were supplied with water and fish stocks naturally by the annual wetland floods. Water balance studies determined the effects of seasonal hydrological patterns on the functioning of the ponds. A trophic model was constructed using Ecopath to evaluate the agroecosystem performance in terms of nutrient flows. The importance of household activities (including Fingerponds) was determined using the sustainable livelihood approach (SLA). Economic analysis determined the performance of fingerponds vis à vis other farming system activities.

Annual floods provided adequate stocks of fish (≥ 3 fish/m^2). Water supply depended on natural processes and in a good year the functional period of the ponds is about 5–6 months. A Fingerpond of about 200 m^2 provided an additional per capita fish supply of 3.4 kg to a household of 7 people. The potential protein supply is about 200 kg/ha, higher than most other farming activities. This can be increased if pond management is further improved. The wetlands support biomass production activities that are nutritionally and financially important to the households. Trophic and economic assessments revealed that Fingerponds promoted nutrient recycling, enhanced food security and increased the gross margin of an average household by 11%. With discounted fixed costs, the return to household labour per person day was 12.49 euros/ha. There was no evidence of negative impacts on the wetland environment in terms of eutrophication of groundwater.

Wetlands resources dominate the livelihood assets of many riparian households. To prevent further encroachment upon them, the productivity of farming needs to be increased to add value to the existing biomass harvesting. Fingerponds contribute to this through their high protein supply per hectare and increased overall yields. This has a potential significance particularly in a region beset by poverty, hunger and malnutrition. The dependence on natural wetland processes (flooding, fish stocking) makes them economically attractive but present uncertainty and high spatio-temporal variability. The investment needed for pond construction may limit adoption by poor households. Institutional support, particularly from the government, NGOs and other local community support groups is required. Further development requires institutional collaboration through multi-stakeholder partnerships; participatory research on integration into existing farming systems, up-scaling and technology improvement; and translation of research results into wetland policies with clear guidelines for communities and decision-makers.

Chapter
1

General introduction

Introduction

This thesis investigates the potential of integrating aquaculture ponds (Fingerponds) into smallholder riparian farming systems at the shores of Lake Victoria, Kenya. The overall aim is to contribute to a strategy for the wise use of the wetlands, i.e., to enhance the existing wetland fishery and seasonal agriculture without compromising ecosystem integrity (Davis, 1993). If successful, this would increase the overall value of the wetland by increasing the use values (fish, crops) while maintaining the non-use values (e.g., biodiversity). This introductory chapter provides a background of wetland uses and values, resource trends, integrated farming system concepts, and introduces the study sites and setting of the experimental Fingerpond systems in Kenya.

Natural wetlands and uses

Wetlands include all forms of landscape characterized by wetness. A number of definitions have been put forward in an attempt to delineate these ecosystems more precisely. The Ramsar Convention on Wetlands of 1971 defines wetlands as *"areas of marsh, fen, peat land or water, whether natural or artificial, with water that is static, brackish or salt including areas of marine water, the depth of which at low tide does not exceed 6 metres"*. This definition encompasses a wide range of diverse landscapes whereby three inherent components of a wetland are manifested: water, hydric soils and hydrophytic vegetation. Wetlands are often the zone of transition between dry land and a water body (lake, river or sea). Often, they support high primary and secondary productivity and biodiversity.

Wetlands have multiple functions and attributes representing values to humanity (Chabwela, 1992; Denny, 1995; Rogeri, 1995; Kairu, 2001). They not only support a rich diversity of flora and fauna but also support the human populations living around them through provision of goods and services. Wetlands are indeed the lifeline for the riparian communities (Silvius et al., 2000). In many developing countries, including those in sub-Saharan Africa, living around a wetland is considered a blessing. Despite being associated with endemic diseases such as malaria and schistosomiasis and livestock parasites such as liver flukes, wetlands provide food for the people and pasture for the livestock during periods of scarcity. For decades, communities at the edges of natural wetlands have cultivated vegetables to meet the household's demands during the dry season and to augment cash income through market sales. Livestock graze along the wetland margins. Table

1.1 shows the economic values of some African wetlands and their contribution to agriculture and fish production. Estimates from West and Central Africa show that the main rivers and floodplains in the region produce an annual fish catch of about 570,000 tonnes and provide employment for about 0.5 million people (Béné, 2005). The diversity of resources explains why Lake Victoria and its catchment can support over 30 million people (Bugenyi, 2001).

Table 1.1: Examples of wetland economic values in Africa and the percentage contribution of agriculture and fish production to the total economic value

	Wetland name			
Attribute	Nakivubo	Hadejia Jama'are	Lake Chilwa	Zambezi basin
Country/region	Uganda	Nigeria	Malawi	Southern Africa
Area (km^2)	5.29	3500	2400	29820
Total Value (million US$ per year)	1.09	16.14	20.99	201.62
Agriculture (%)	5.5	68.2	5.7	24.8
Fish (%)	0.3	21.7	89.1	39.0

Adapted from Schuyt, 2005

Some common resources tied to traditional functions of wetlands around Lake Victoria are shown in Figure 1.1. Attempts have been made to value the services and functions of wetlands economically to express the importance of wetlands in support of policy and decision-making (Barbier et al., 1997; Emerton at al., 1999; Stuip et al., 2002; Ramsar Convention Secretariat, 2004). The non-use (e.g., biodiversity, cultural or religious) values of wetlands are more difficult to quantify in monetary terms but are very important to local communities. As an example, the river Nyando and the surrounding lowland on the eastern shores of Lake Victoria, Kenya, are said to provide a habitat for a legendary python of good fortune associated with good rains and high yields.

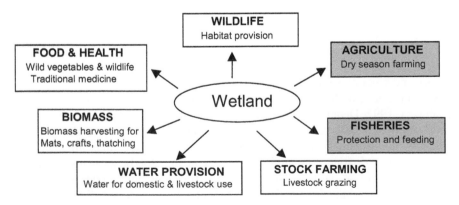

Figure 1.1: Some common functions of African wetlands (the shaded shows the area of interest of this study)

The Lake Victoria wetlands

Lake Victoria, East Africa, the second largest freshwater lake in the world, is surrounded by vast wetlands. The lake has a total surface area of 68,800 km^2 of which 49%, 45% and 6 % belong to Tanzania, Uganda and Kenya, respectively. It has a shoreline of 3,440 km and is characterized by both littoral and floodplain wetlands. In Kenya, some of the most important wetlands occur in the floodplains of the major rivers flowing to Lake Victoria. These include the Yala, Nyando, Sondu-Miriu, Nzoia and Gucha rivers (Figure 1.2). These wetlands form an integral part of the rural economies and livelihood of the local people.

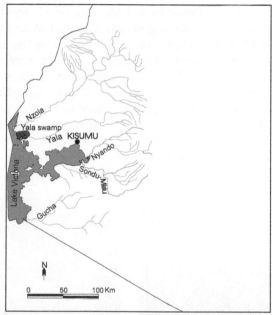

Figure 1.2: The Lake Victoria Kenyan side: shoreline and major rivers and floodplains

Threats to wetlands

The diversity and importance of wetland functions to riparian communities has, ironically, often resulted in threats to the existence of these ecosystems. Over-exploitation has led to the destruction of large areas of wetlands worldwide. It is estimated that 50% of the global wetlands have been lost since 1900 (Moser et al., 1996). In Africa, the actual loss is unknown. Whilst local communities can narrate the trends of decline in wetland resources in their home regions, there is a lack of sufficient documented evidence of wetland destruction. This gap in knowledge, combined with the absence of clear policies and regulations on wetland conservation, protection and wise use, causes wetlands to remain a very vulnerable resource in many African countries.

Trends in the wetland resources of Lake Victoria, Kenya

Overfishing and wetland degradation have led to a decline of subsistence and open-water fishery yields in East African freshwater bodies, including Lake Victoria (Okeyo-Owuor, 1999; Odada et al., 2004). Other factors that may have contributed to the current problems in the lake include eutrophication (Hecky, 1993); destructive fishing practices (Bwathondi, Ogutu-Ohwayo and Ogari, 2001, cited in Odada et al., 2004) and threats to its biological diversity, e.g., by introduction of exotic species like the Nile perch and the water hyacinth (Hall and Mills, 2000; Goudswaard et al., 2002; Balirwa et al., 2003; Aloo, 2003). The changes have resulted in poverty and protein deficiency among the local riparian communities who used to depend on the traditional littoral fishery for generations. Most of the fish currently harvested from the open water is destined to fish-processing factories and the export market, thus creating unfavourable competition with local markets. At the same time, the seasonal post-flood wetland fishery has declined significantly (Mr. M. Onyango, Kusa villager, pers. comm.). Ochumba and Manyala (1992) reported that the fish yields in the Sondu Miriu River draining to Lake Victoria had dropped to 108 tonnes from 668 tonnes in 1959. The reliance on capture fishery by the people around the lake seems to be no longer feasible. Population growth and increased poverty have contributed to encroachment and extensive conversion of the wetland emergent macrophyte zone for seasonal crop production (Figure 1.3). Harvesting of the wetland natural biomass (e.g., papyrus) has increased and is now an essential part of people's livelihoods. This often results in over-harvesting and destruction of the natural wetland vegetation. Evidently, there is need for enhanced, sustainable production from wetlands to ensure food security while at the same time maintaining the non-use functions of the wetlands.

Figure 1.3: Extensive conversion of emergent macrophyte zone into seasonal crop production

Wetland fishery potential

African inland wetlands, particularly floodplains, are vital for the continent's fishery (Welcomme, 1975). However, the contribution of the wetland fishery to household economies is not well documented as it is characterized by seasonal variability and is mostly at the subsistence level: people harvest fish from permanent or seasonal pools or from the littoral wetlands shortly after the floods. Catches are rarely recorded and do not appear in official catch statistics. Some fish are consumed immediately within households while the surplus may be sold for income. Studies from Bangladesh indicate that the wetland fishery, especially in the floodplains, contributes significantly to the livelihoods of riparian communities (Craig et al., 2004). Ricefield fish catches in countries like Cambodia, Malaysia and Thailand can yield between 50 and 300 kg per hectare per year (Gregory and Guttman, 2002).

Food security in sub-Saharan African; trends and the potential contribution from aquaculture

In the 2000-2002 period, 24% of the world's undernourished population resided in sub-Sahara Africa (FAO, 2004a). About 50% consisted of smallholder farmers. In many situations where food security is precarious, women and children are the most vulnerable. The target of the World Food summit in 1996 was to reduce the undernourished population from 792 million to 400 million by 2030. For this to be achieved, more effort should be directed to low-cost and sustainable production systems. Fish is nutritionally important and provides 20% or more of animal proteins to the majority of the population in sub-Sahara Africa (FAO, 2004b). However, the per capita supply has declined over the last decade. Earlier studies by Kapetsky (1994) indicated that there is a high biophysical potential for aquaculture in the continent.

During the 1950s and 1960s, numerous efforts were directed to the development of conventional aquaculture systems (Brummett and Williams, 2000). However, this was not very successful. Although the continent tried to keep pace with the average annual world growth rate of 8.9 % in aquaculture since 1970, the contribution is still low compared to capture fisheries (FAO, 1997; 2004a). Generally, the growth of aquaculture in Africa has not matched its potential (Figure 1.4). Machena and Moehl (2001) identified some of the limitations in the development of aquaculture in sub-Saharan Africa: inadequate inputs (especially supplemental feeds and fingerlings), inappropriate technologies, and weak research and extension. Despite these limitations, aquaculture is still considered to have great potential for rural economies (Vincke, 1995; Brummett and Williams, 2000). Halwart and van Dam (2006) argue that the overall production and livelihood security can be increased by integrating fish production with small-scale irrigation. Clearly, there is a need to explore the integration of aquaculture into other farming systems.

The Lake Victoria wetlands and scattered swamps around other large African water-bodies present a great potential for aquaculture development (Bernacsek, 1992). If only 1% of the 12 million hectares of Africa's floodplains could be utilized for aquaculture using appropriate intermediate techniques, over 100,000 tonnes per year of fish would be realized (Balarin 1988 in COFAD, 2002). Using the estimates of annual production from FAO (2004 a), this is equivalent to about a 5% increase.

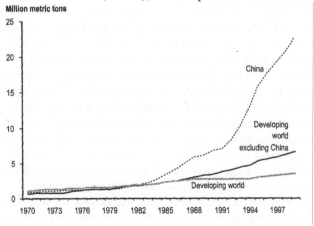

Figure 1.4: Trends in total aquaculture production, 1970-1999. (Based on Delgado et al., 2003)

Integrated farming systems: focus on aquaculture
Integrated agriculture-livestock production systems are an old tradition in Africa, but integrated agriculture-aquaculture systems are a relatively new concept although they have been practiced for a long time in Asia (Devendra, 1995; Kangmin and Peizhein, 1995; Tokrisna, 1995; Prein, 2002). Integration of aquaculture into other agricultural activities presents a promising opportunity as it enhances synergy and minimizes the risk associated with single enterprises. While such systems have proved successful in Asia (Mukherjee, 1995; Haylor and Bhutta, 1997; Fernando and Halwart, 2000), the challenge remains to adapt them to the African situation (Brummett, 1999). What is required is knowledge on the design of integrated aquaculture systems and on what makes them ecologically and economically sustainable in relation to the context in which they are to be developed. It is also necessary to understand the determinants of adoption of new technology (Pullin and Prein, 1995). This study addresses some of these issues for the development of wetland-based integrated aquaculture-agriculture systems (Fingerponds) at the shores of Lake Victoria in Kenya.

Fingerponds
Fingerponds are integrated fish and crop production systems. It is an innovative, semi-intensive technology aimed at enhancing wetland products based on the wetland's natural functions (Denny, 1989). Fingerponds may be regarded as enhancement of the traditional fishery whereby local knowledge on the flood pool fishery is developed to meet the increased demand for fish proteins from the villages adjacent to natural wetlands (COFAD, 2002). They are earthen ponds excavated in the fringe wetlands during the dry season: the excavated soil is spread around the ponds to create raised-bed gardens for vegetable production.

The ponds resemble the natural flood pools traditionally used for wetland fish capture by local communities while the gardens are a continuation of the existing seasonal swamp margin vegetable patches. They are called "Fingerponds" because, from a bird's eye view, several of these narrow channel-like ponds appear like "fingers" penetrating the emergent macrophyte zone. The fish are trapped in the ponds during flood recession and manure and vegetable wastes from the adjacent village are used to improve pond productivity. Locally-demanded vegetables are grown on the raised beds. The advantage of this system is that it enhances diversity of produce as well as synergy between different components of the farming system. Pond water may be used to irrigate the gardens while the sludge from the pond bottom is removed during the dry season and spread over the raised beds as a fertilizer. The excess vegetables from the adjacent gardens can be chopped and used as fish food or composted and applied as green manure. The Fingerponds concept is similar to the Chinese dike pond systems (Korn, 1996), Mexican 'hortillonages' (Micha et al., 1992), and agri-piscicultural systems developed in Rwanda (Barbier et al., 1985) where crop and fish production systems are integrated. However, the system is unique in that water is not regulated, relying on natural flooding of the wetland during the rainy season to supply water and stock the ponds with fish.

Figure 1.5 shows the design and appearance of the experimental Fingerponds. For this study, ponds of 192 m^2 with a sloping bottom of 1m at the shallow end to 2 m at the deep end and an adjacent garden of similar area were constructed. Fingerponds provide an option for enhancing benefits for the local population without engaging in drainage of wetlands for agriculture, or other forms of unsustainable wetland utilization. They can be integrated into existing farming systems and hence provide additional protein to meet the nutritional deficiency among the poor rural communities.

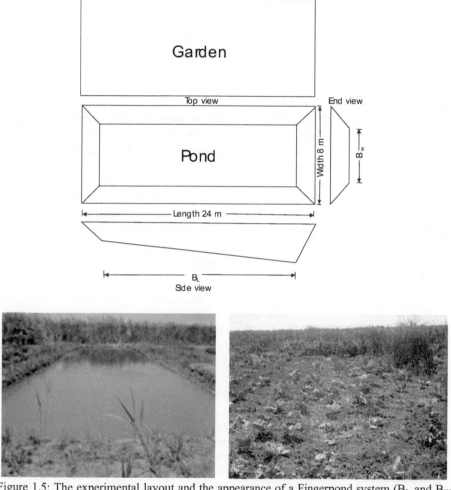

Figure 1.5: The experimental layout and the appearance of a Fingerpond system (B_L and B_W are pond bottom length and width respectively). Photos show the pond (left) and the garden (right)

Objectives of the study

The Fingerponds concept is linked to the wetland and the terrestrial ecosystems. It incorporates two existing wetland functions, the flood-pool fishery and seasonal agriculture, into the terrestrial farming systems thereby creating potential synergies and benefits. In this study I explored how Fingerponds might be integrated successfully into the existing riparian farming systems.

The specific objectives were;

1. To assess the current uses and values of natural wetlands and the biophysical suitability of Fingerponds in East African freshwater wetland ecosystems;
2. To evaluate the fish production potential of Fingerpond systems by simple manipulations of fertilization with livestock manure;
3. To evaluate the ecological performance and sustainability of Fingerponds;
4. To evaluate the contribution of Fingerponds to riparian peoples' livelihoods;
5. To assess the potential environmental effects of Fingerponds on natural wetlands in terms of potential eutrophication, effects on aquatic macrophyte biomass and species composition.

Summary of the study approach

The general setting of this study is aimed at understanding the functioning of Fingerponds in the context of the entire riparian farming system. A multi-disciplinary approach is used to unravel the functioning of these systems. Figure 1.6 shows the context and the approach of the study. The core components of the study are the Fingerponds: the wetland aquaculture ponds with their associated gardens. The context for Fingerponds is the interface between the terrestrial farming systems and the wetland activities. The study takes a systems approach, ranging from the biophysical (water, soil, fish, inputs) to the socio-economic aspects (costs, benefits, livelihoods). This combination of ecological and socio-economic analyses leads to an assessment of the performance and sustainability of the overall scheme.

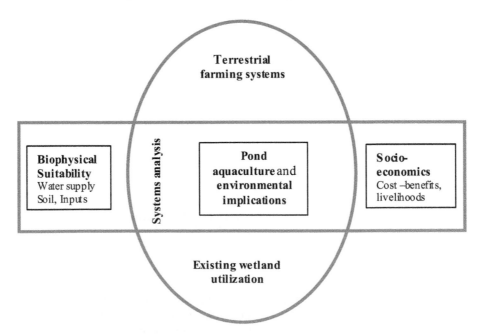

Figure 1.6: Schematic overview of the study approach

The location of the Fingerpond experimental sites

The study was conducted between 2002 and 2005 in association with the European Union-supported Fingerponds Project in partnership with three East African countries: Kenya, Uganda and Tanzania. In Kenya, the ponds were located in two sites near the shores of Lake Victoria (Figure 1.7). The Nyangera site is on the northern shores of the lake in the littoral wetlands sandwiched between Kadimu and Usenge Bay (S 0° 3' 55.9", E 34° 4' 52.2") while the Kusa site is on the eastern shores of the lake bordering Nyakach bay (S 0° 18' 1.2", E 34° 53' 21.3"). In Nyangera, Fingerponds were constructed about 500 m from the shoreline in the emergent macrophyte zone. The vegetation is composed of mixed stands of emergent macrophytes dominated by *Phragmites* sp., *Typha domingensis* and *Cyperus papyrus*. In Kusa, Fingerponds were constructed about 4 km from the lake shoreline at the periphery of the floodplain wetland. The wetland ecosystem is dominated by papyrus with *Vossia cuspidata* occurring mainly on the river banks. Small patches of isolated stands of *Phragmites* sp. and *Typha domingensis* are also common. At the wetland margin adjacent to the terrestrial ecosystem, the vegetation consist of mainly *Cyperus* spp. and *Cynodon dactylon*. There were four other experimental sites in East Africa; two in Uganda at the shores of Lake Victoria and two in Tanzania in the Rufiji floodplain (Denny et al., 2006).

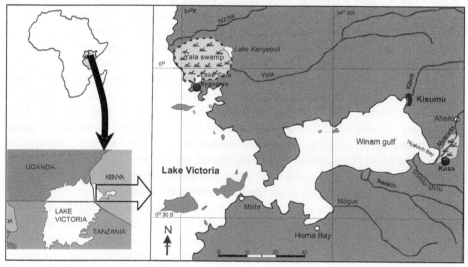

Figure 1.7. Fingerponds study sites in Kenya

Communities at the study sites

At each Fingerponds site, the local community was involved closely in site selection, construction, experimental setup and co-management.

Nyangera - the Nyangera school community

Nyangera primary school is found in Usigu division, Bondo district in Nyanza province of Kenya. This is a public day school founded in the 1950s. In 2003, the school population was 330 pupils. This includes both the Early Childhood Development (ECD) section and primary school. The school children come from the local villages where they live with their parents or guardians. The school community includes 11 teachers. About 30% of the pupils have single parent families or are orphans (Mrs. D. Ngoye, pers. com). The pupils from the ECD section and the orphans are provided with food under the school feeding programme. The school cultivates vegetables at the Lake Victoria wetland margin to augment the food requirements for this programme. Fingerponds were introduced and incorporated into the school wetland farming activities.

Land in the surrounding area is used mainly for subsistence agriculture, livestock grazing and agroforestry. The main livelihood activities for the local people are fish trading, seasonal swamp agriculture, subsistence farming, formal employment, livestock farming and bicycle transport (popularly known as *boda boda*). Seasonal wetland farming is common in littoral wetlands especially during the dry season. Among the crops commonly grown are kales, spinach, cowpeas, sweet potatoes, tomatoes, cassava, banana, arrow roots and sugar cane. The expanding population has led to increased demand for seasonal agricultural land from the wetland. Hunting for wild game is common during the dry season. According to the local people, the main regular flood occurs between April and June. The extent of floods is determined by the amount of precipitation in the catchment.

Kusa - the Komolo women group, Kayano village in Kusa

Kayano village is located in Kusa in Lower Nyakach division of Nyando district in Nyanza province, Kenya. Nyakach is well-known for annual floods, especially on areas adjacent to the River Nyando. The local community is organized into villages that normally consist of one or two clans. Most households are made up of extended families. Land ownership by households is mainly through paternal inheritance. Household land boundaries are rarely fenced but are demarcated by live fences of sisal or euphorbia. The wetlands belong to the government, however, the local people are allowed to utilize the resource for their livelihood. In Kusa the group participating in the Fingerponds project consisted of women from 12 households living adjacent to the wetland. After Fingerpond construction the men were mainly involved in papyrus harvesting and other non-farm income activities, while the women were mostly involved in on-farm activities and Fingerponds.

The Kusa landscape is characterized by two distinct zones: the highland or hilly region and the lowland bordering the lake and Nyando wetland. In the terrestrial ecosystem the land is used for subsistence agriculture and smallholder animal husbandry. The main crops grown in the terrestrial farming systems are maize, finger millet, groundnuts, cotton, bananas, sugarcane and vegetables. Animal production is small-scale whereby cattle, sheep, goats and poultry are reared for subsistence.

The natural wetland ecosystem is an important resource for the local community. After clearing the natural wetland vegetation, seasonal crops, mainly beans, maize, vegetables (kales, cowpeas and tomatoes) and arrowroots are grown during the dry season, along with some biennial crops (sugarcane and bananas). Natural products that are harvested from the wetlands include papyrus culms for mat making; and fish, after flood recession. According to the local people, the regular flood occurs in March/April and depends on the amount of precipitation in the river Nyando catchment in the highland regions of Nandi and Kericho districts. Flood recession farming especially for vegetables start at the end of May to June. Crop cultivation ends just before the next floods. Some crops such as arrowroots, bananas, sugarcane can tolerate short flooding as long as the plant leaves are not covered by water. Some terrestrial, vegetatively propagated crops such as cassava and sweet potatoes are transferred to the wetland during the dry season and are returned back just before the floods. Thus, the wetland acts as a seed-bank for the local community.

The importance of fish to the local households

Fish is traditionally an important component of the daily diet among the Luo community. However, due to the decline in fishery in Lake Victoria and surrounding wetlands, consumption has dropped to only a few meals per week. Prices are determined by the market forces created by the fish processing plants, and by demand from urban centres in the region. Because of the demand-driven high prices, the affordability for the poor households has declined.

The source of protein in many rural African communities is milk. However, dairy farming around Lake Victoria has not been very successful. Tsetse flies affect particularly exotic cattle, so the dominant cattle breeds are the low-yielding traditional East African zebu. This leaves the people vulnerable to protein deficiency. The poorer members of the community, who constitute the majority, have to rely on the less favoured sardine-like fish *Rastrineobola argentia* popularly known as "dagaa", and fillet-stripped fish popularly known as *"mkongo wazi"* or bare backbone left-overs from the fish processing plants (Mr. Mathews Onyango, pers comm.).

The Kusa/Nyando wetland: its wildlife diversity and mysteries

The Nyando wetland is about 3000 hectares and is characterized by high wildlife diversity. Large mammals such as hippopotamus are found at the river mouth and occasionally wander into the margin of the wetland and even into the village at night. Crocodiles are found on the river banks. Occasionally, people have been attacked and killed by the wildlife. There is also a wide diversity of wetland birds such as herons, crested cranes, hamerkops, egrets, swamp fly catchers, kingfishers etc. The common fish species in the wetland include lungfish (*Protopterus aethiopicus*), mudfish, (*Clarias* sp), and, *Schilbe intermedius*.

One famous story about the Nyando wetland is the presence of a wetland python locally known as "omieri". This python is believed to bring blessing to the local community whenever it visits the adjacent village. The python (mostly "she") chooses an unsuspecting household, where it lays eggs and broods. Omieri is believed to be harmless and retreats back to the wetland once the eggs hatch. Such visits are rare and bring a lot of excitement whenever they occur. During the visits, the python is fed with all sorts of delicacies ranging from chicken to the famous lakeside dish "ugali". The local community believes that such visits are followed by a bumper harvest in the terrestrial rain-fed agriculture, so the reptile has to be treated well. The visits have existed for generations and can be remembered as far back as 1948. However, the most memorable visits are said to have occurred in 1964 and 1965. In 1987 the python appeared in one of the villages around the Nyando wetland. Unfortunately, it was scalded by a bush fire and was flown to Nairobi National Museum's Snake Park for treatment amidst an outcry from the local residents. Some argued that their luck was being taken away. The python later died and was returned back to the village where it was buried in a coffin with full honours. One old man confided to me that the boys who started the fire perished in mysterious circumstances soon after the death of the serpent. The most recent visits occurred in February 2003 and 2006. There are mixed views about the python. The conservationists from the Kenya Wildlife Services (KWS) argue that the visit to the village and subsequent brooding is part of the reptile's behaviour to give her off-spring a better chance of survival by laying eggs in advance of rains. The community is divided; some Christians do not want to hear anything about the serpent, others choose to keep quiet about it. Nevertheless, many villagers insist that the snake is sacred and is associated with good rains and high agricultural yields. Whichever is true is yet another mystery, however is a good example of an overlap of culture and conservation.

The structure of the thesis

This thesis consists of 8 chapters. Chapter 1 (this chapter) gives an introduction on wetland uses and the potential for integrated wetland aquaculture, the description of the study sites and the approach to the research. Chapters 2 and 3 evaluate the biophysical suitability of Fingerponds in the context of the littoral and floodplain wetlands at Lake Victoria, Kenya. Chapter 4 assesses the aquacultural potential of Fingerponds and the effects of pond management through addition of manure on water and sediment quality and fish yields. In Chapters 5 and 6, the Fingerponds systems are evaluated from a broader farming system perspective. Chapter 5 assesses the Fingerponds from an agro-ecosystem perspective using nutrient flows (nitrogen) whilst Chaper 6 evaluates the systems in terms of contribution to household food security and livelihoods *vis à vis* other farming system activities. The potential ecological and social implications of introducing Fingerponds into wetlands are addressed in Chapter 7. The final chapter (Chapter 8), synthesises the results of integrated aquaculture (Fingerponds) in the context of the existing farming system and draws conclusions.

References

Aloo, P.A., 2003. Biological diversity of the Yala swamp lakes, with special emphasis on fish species composition, in relation to changes in the Lake Victoria basin (Kenya); threats and conservation measures. Biodiversity and Conservation 12, 905-920.

Balarin, J.D., 1988. Aquaculture development planning: The logistics of fish farm project appraisal. In: King, H.R and K.H. Ibrahim (eds.) Village level aquaculture development. Proceedings of Commonwealth consultative workshop, Freetown, Sierra Leone, Commonwealth Secretariat Publications, pp. 92-106.

Balirwa, J.S., Chapman, C.A., Chapman, L.J., Cowx, I.G., Geheb, K., Kaufman, L., Lowe-McConnell, R.H., Seehausen, O., Wanink, J.H., Welcomme, R.L., Witte, F., 2003. Biodiversity and fishery sustainability in the Lake Victoria basin; an unexpected marriage? Bioscience 53(8), 703-716.

Barbier, E.B., Acreman, M.C., Knowler, D., 1997. Economic valuation of wetlands: a guide for policy makers and planners. Ramsar Convention Bureau, Gland, Switzerland.

Barbier, P., Kalimanzira, C., Micha, J-C., 1985. L'aménagement de zones marécageuses en écosystémes agro-piscicoles. Le project Kirarambogo au Rwanda (1980-1985) ed. FUCID, Namur, Belgique 35 pp.

Béné, C., 2005. Contribution of inland fisheries to rural livelihoods and food security in Africa: An overview. In: M.L.Thieme, R. Abell, N. Burgess, E. Dienerstein, B. Lehner, D.Olson, G.G, Teugels, A.K. Toham, M.L.J. Stiassny and P. Skelton (eds.) Island Press, pp. 6-11.

Bernacsek, G.M., 1992. Research priorities in fisheries management as a tool for conservation and rural development. In: E. Malby, P.J. Dugan and J.C. Lefeuvre (eds.). Conservation and Development: The sustainable Use of Wetland Resources. Proceedings of the Third Internationals conference, Rennes, France, 19-23, September 1988. IUCN, Gland, Switzerland, pp. 131-144.

Brummett, R.E., 1999. Integrated aquaculture-agriculture in sub-Saharan Africa. Environment, Development and Sustainability 1, 315-321.

Brummett, R.E., Williams, M. J., 2000. The evolution of aquaculture in African rural and economic development. Ecological Economics 33, 193-203.

Bugenyi, F.W.B., 2001. Tropical freshwater ecotones: their formation, functions and use. Hydrobiologia 458, 33-43.

Bwathondi, P.O.J., Ogutu-Ohwayo, R., Ogari, J., 2001. Lake Victoria fisheries management plan. In: I.G. Cowx and K.Crean (eds.) LVFRP/TECH/01/06, Technical Document No. 16, 64 pp.

Chabwela, H.W., 1992. The exploitation of wetland resources by traditional communities in the Kafue Flats and Bangweulu basin. In: E. Malby, P.J. Dugan and J.C. Lefeuvre (eds.). Conservation and Development: The sustainable use of wetland resources. Proceedings of the Third Internationals Conference, Rennes, France, 19-23, September 1988. IUCN, Gland, Switzerland, pp. 31-39.

COFAD, 2002. Back to basics; traditional inland fisheries management and enhancement systems in Sub-Saharan Africa and their potential for development, Eschborn, Germany. 203 pp.

Craig, J.F., Hall, A.S., Barr, J.J.F., Bean, C.W., 2004. The Bangladesh floodplain fisheries. Fisheries Research 66, 271-286.

Davis, T.J. (ed.)., 1993. Towards the wise use of wetlands. Wise use project. Ramsar Convention Bureau, Gland Switzerland.

Delgado, C.L., Wada, N., Rosegrant, M.W., Meijer, S., Ahmed, M., 2003. Fish to 2020: Supply and demand in changing global markets. International Food Policy Research Institute, Washington and Worldfish Centre, Malaysia, 226 pp.

Denny, P., 1989. Wetlands. In: Strategic resources planning in Uganda. UNEP Report IX, 103 pp.

Denny, P., 1995. Benefits and priorities for wetland conservation. The case for national wetland conservation strategies. In: Wetland Archaeology and Nature conservation, M. Cox, V. Straker & D Taylor (eds.). Proceedings of international conference on wetland archaeology and nature conservation, University of Bristol, 11-14 April, 1994. HMSO, UK.

Denny, P., Kipkemboi, J., Kaggwa, R., Lamtane, H., 2006. The potential of Fingerpond systems to increase food production from wetlands in Africa. International Journal of Ecology and Environmental Sciences, 32(1), 41-47.

Devendra, C., 1995. Integration of agriculture and fish farming in Indonesia. In: J.-J. Symoens and J.-C. Micha (eds.) Seminar "The management of integrated freshwater agro-piscicultural ecosystems in tropical areas", Brussels, 16-19, May 1994. Technical Centre for Agricultural and Rural Co-operation (CTA) Royal Academy of overseas sciences, Brussels. pp. 329-341.

Emerton, L., Lyango, L., Luwum, P., Malinga, A., 1999. The present economic value of Nakivubo urban wetland, Uganda, National Wetlands Programme and IUCN.

FAO, 1997. Review of the state of the world aquaculture. FAO Fisheries Circular No. 886. FIRI/C 886 (Rev. 1), Food and Agriculture Organisation of the United Nations, 163 pp.

FAO, 2004 a. State of World Fisheries and Aquaculture (SOFIA). Food and Agriculture Organization of the United Nations, Fisheries Department, Rome, 153 pp.

FAO, 2004 b. Aquaculture extension in sub-Saharan Africa. Fisheries Circular No. 1002. Food and Agriculture Organization of the United Nations, Rome 55 pp.

Fernando, C.H., Halwart, M. 2000. Possibilities for the integration of fish farming into irrigation systems. Fisheries Management and Ecology 7, 45-54.

Goudswaard, P.C., Witte, F., Katunzi, E.F.B., 2002. The tilapiine fish stock of Lake Victoria before and after the Nile perch upsurge. Journal of Fish Biology 60, 838-856.

Gregory, R., H. Guttman., 2002. The rice field catch and rural food security. Chapter 1, In: P. Edwards, D.C. Little and H Demaine (eds.) Rural aquaculture. CABI Publishing, Oxon, UK.

Hall, S.R., Mills, E.l., 2000. Exotic species in large lakes of the world. Ecosystem Health and Management 3, 105-135

Halwart, M., van Dam, A.A. (eds.), 2006. Integrated irrigation and aquaculture in West Africa: concepts, practices and potential. Food and Agriculture Organization of the United Nations (FAO), Rome 181 pp.

Haylor, G., Bhutta, M.S., 1997. The role of aquaculture in the sustainable development of irrigated farming systems in Punjab, Pakistan. Aquaculture Research 28, 691-705.

Hecky, R.E., 1993. The eutrophication of Lake Victoria. Verh. Internat. Verhein. Limnol. 25, 39-48.

Kairu, J.K., 2001. Wetland use and impact on Lake Victoria, Kenya region. Lakes and Reservoirs: Research and Management 6, 117-125.

Kangmin, L., Pehzein, L., 1995. Integration of agriculture, livestock and fish farming in the Wuxi region of China In: J.-J. Symoens and J.-C. Micha (eds.) Seminar "The management of integrated freshwater agro-piscicultural ecosystems in tropical areas", Brussels, 16-19, May 1994. Technical Centre for Agricultural and Rural Co-operation (CTA) Royal Academy of overseas sciences, Brussels. pp. 309-328.

Kapetsky, J.M., 1994. A strategic assessment of warm-water fish farming potential in Africa. CIFA Technical Paper, No. 27. Rome, FAO, 67 pp.

Korn, M. 1996. The Dike-Pond concept: sustainable agriculture and nutrient recycling in China. Ambio 25 (1), 6-13.

Machena, C., Moehl, J., 2001. Sub-Saharan African aquaculture: regional summary. In: R.P. Subasinghe, P. Buana, M.J. Phillips, C. Hough, S.E. McGladdery and J.R. Arthur (eds.). Aquaculture in the Third Millennium. Technical Proceedings of the conference on aquaculture in the Third Millennium, Bangkok, Thailand, 20-25 February, 2000, NACA, Bangkok and FAO, Rome, pp. 341-355.

Micha. J.-C., Hallen, H., Rosado Couch, et J.-L., 1992. Changing Tropical Marshlands into agro-piscicultural systems. In: E. Malby, P.J. Dugan and J.C. Lefeuvre (eds.). Conservation and Development: The sustainable Use of Wetland Resources. Proceedings of the Third Internationals conference, Rennes, France, 19-23, September 1988. IUCN, Gland, Switzerland. xii, pp. 179-186.

Moser, M., Prentice, C., Frazier, S., 1996. A global overview of wetland loss and degradation. Paper presented at the 6th meeting of the conference of the contacting parties in Brisbane, Australia, 19th-27th March 1996.

Mukherjee, T.K., 1995. Integrated crop-livestock-fish production systems for maximizing productivity and economic efficiency of smallholder's farms. In: J.-J. Symoens and J.-C. Micha (eds.), Seminar "The management of integrated freshwater agro-piscicultural ecosystems in tropical areas", Brussels, 16-19, May 1994, Technical Centre for Agricultural and Rural Co-operation (CTA) Royal Academy of Overseas Sciences, Brussels, pp. 121-143.

Ochumba, P.B.O., Manyala, J.O., 1992. Distribution of fishes along the Sondu-Miriu River of Lake Victoria, Kenya with special reference to upstream migration, biology and yield. Aquaculture and Fisheries Management, Oxford Vol.23, No 6, pp 701-719.

Odada, E.O., Olago, D.O., Kulindwa, K., Ntiba, M., Wandiga, S., 2004. Mitigation of environmental problems in Lake Victoria, East Africa; Causal chain and policy options analyses. Ambio 33(1-2), 13-23.

Okeyo-Owuor, J.B., 1999. A review of biodiversity and socio-economics research to fisheries in Lake Victoria. IUCN, 48 pp.

Prein, M., 2002. Integration of aquaculture into crop-animal systems in Asia. Agricultural Systems 71, 127-146.

Pullin, R.S.V., Prein, M., 1995. Fishponds facilitate natural resources management on small-scale farms in tropical developing countries. In: J.-J. Symoens and J.-C. Micha (eds.) Seminar "The management of integrated freshwater agro-piscicultural ecosystems in tropical areas", Brussels, 16-19, May 1994. Technical Centre for Agricultural and Rural Co-operation (CTA) Royal Academy of overseas sciences, Brussels, pp. 169-186.

Ramsar Convention Secretariat, 2004. Ramsar handout for wise use of wetlands. 2nd edition, Ramsar Secretariat, Gland Switzerland

Rogeri, H., 1995. Tropical freshwater wetlands: A guide to current knowledge and sustainable management. Kluwer Academic Publishers, Dordrecht, 349 pp.

Schuyt, K.D., 2005. Economic consequences of wetland degradation for local populations in Africa. Ecological Economics 53, 177-190.

Silvius, M.J., Oneka, M., Verhagen, A., 2000. Wetlands: lifeline for people at the edge. Phys. Chem. Earth (B), 25 (7-8), 645-652.

Stuip, M.A.M., Baker, C.J., Oosterberg, W., 2002. The socioeconomics of wetlands, Wetlands International and RIZA, The Netherlands.

Tokrisna, R., 1995. Integration of agriculture, livestock and fish farming in Thailand. In, J.-J. Symoens and J.-C. Micha (eds.), Seminar "The management of integrated freshwater agro-piscicultural ecosystems in tropical areas", Brussels, 16-19, May 1994, Technical Centre for Agricultural and Rural Co-operation (CTA) Royal Academy of overseas sciences, Brussels, pp. 121-143.

Vincke, M.M.J., 1995. The present state of development in continental aquaculture in Africa. In: J.-J. Symoens and J.-C. Micha (eds.) Seminar "The management of integrated freshwater agro-piscicultural ecosystems in tropical areas", Brussels, 16-19, May 1994. Technical Centre for Agricultural and Rural Co-operation (CTA) Royal Academy of overseas sciences, Brussels, pp. 27-61.

Welcomme, R.L. 1975. The fisheries ecology of African floodplains. CIFA Technical Paper 3, 54 pp.

Chapter

2

Biophysical suitability of smallholder integrated aquaculture-agriculture systems (Fingerponds) in East Africa's Lake Victoria freshwater wetlands

Abstract

Most riparian communities living along the shores of Lake Victoria rely on wetland farming or harvesting of natural wetland products for their livelihoods. The potential for the enhancement of wetland benefits through smallholder aquaculture systems integrated into existing farming activities was investigated. In two experimental sites near Lake Victoria, Kenya, ponds were dug in wetlands and were used for fish production whilst excavated soil was used to create raised bed gardens for vegetable production. These integrated fish/crop production systems are called 'Fingerponds'. Annual floods stocked the ponds naturally. After flood recession, manure from the adjacent village was used to improve pond productivity. Locally demanded vegetables were grown in the gardens.

The predominantly clayey soils around Lake Victoria are generally suitable for aquaculture. The pilot study revealed that earthen ponds dug in the wetland (Fingerponds) can be adequately stocked during annual floods with local fish species (≥ 3 fish/m^2). The dominant fish are three species of tilapia (*Oreochromis niloticus, O. leucostictus* and *O. variabilis*), *Clarias* sp. and *Protopterus* sp. Manure for pond fertilization is adequately available from the local villages. Fingerponds have the potential of enhancing the existing wetland benefits through fish and vegetable production.

Key words: Smallholder integrated aquaculture-agriculture, freshwater wetlands, Lake Victoria, Fingerponds, riparian communities

Publication based on this chapter:

Kipkemboi, J., van Dam, A.A., Denny P., 2006. Biophysical suitability of smallholder integrated aquaculture-agriculture systems (Fingerponds) in East Africa's Lake Victoria freshwater wetlands. International Journal of Ecology and Environmental Sciences 32(1), 75-83.

Introduction

Wetlands are an important resource for the livelihoods of riparian communities (Silvius et al., 2000). By virtue of their relatively high productivity, these ecosystems can support endemic wildlife and a considerable human population living around them. For rural communities at the shores of Lake Victoria, Kenya, natural wetlands provide a variety of natural products ranging from papyrus biomass which has multiple uses, to food products such as fish and seasonal crops.

Tropical wetlands are known to be very productive (Denny, 1985; Ellenbroek, 1987). However the human population explosion, particularly in sub-Saharan Africa, coupled with unsustainable exploitation has led to a decline in wetland goods, particularly fisheries (Balirwa, 1998). This is evidenced by poverty among the riparian communities as well as unsustainable encroachment upon wetland ecosystems. For wetlands to continue supporting communities at the edge of the swamps, effective utilization and management techniques have to be put in place. More efforts should be directed to enhancing existing wetland uses rather than complete alteration of ecosystem functions or new uses (Symoens, 1995).

Agriculture-livestock production systems are an old tradition in Africa. Integrated aquaculture-agriculture systems are a relatively new concept in Africa but have been practiced for a long time in Asia (Devendra, 1995; Kangmin and Peizhein, 1995; Haylor and Bhutta, 1997; FAO, 2000; Fernando and Halwart, 2000; Prein 2002). The main benefit of such systems is synergy between farming system components and reduction of risks arising from dependence on one enterprise.

African aquacultural potential has not been exploited fully but has a promising future for rural economies (Vincke, 1995; Brummett and Williams, 2000). Earlier studies have indicated that there is a high biophysical potential for aquaculture in the continent (Kapetsky, 1994; 1995), yet there have been limited ventures into this enterprise despite the fact that it can provide an alternative source of food (protein) as well as income. Some of the constraints which may have restricted the development of aquaculture include water requirements, soil properties, inputs, market for products and infrastructure (Aguilar-Manjarez and Nath, 1998). Furthermore, the history of failures of the initial donor-funded aquaculture projects discouraged adoption of aquaculture. Many attempts aimed at developing aquaculture have emphasized intensive high-input systems rather than enhancing and intensifying traditional fish production techniques (COFAD, 2002). There is still a dearth of information on how to design integrated aquacultural systems with sound ecological and economic sustainability, particularly in Africa (Pullin and Prein, 1995).

Fingerponds are integrated fish and crop production systems. They are an innovative, semi-intensive technology aimed at enhancing wetland products based on natural wetland functioning. A proposal for such intermediate technologies was made by Denny (1989). These systems are based on existing wetland services specifically fisheries and agriculture. Ponds are dug into the wetlands and used for fish production while soils excavated are used to create raised bed gardens between the ponds (Figure 2.1). The ponds are naturally stocked by floods. After flood recession, manure from local villages is used to improve pond productivity. Locally demanded vegetables are grown on the raised beds. Fingerponds provide benefits for the local population without necessarily engaging in massive draining for agriculture or filling for human settlement and industrial developments which is destructive to

Figure 2.1: A Fingerpond in a natural wetland, Nyangera, Kenya

the wetlands. One of the main expected benefits is protein supply to relieve protein deficiency among the poor rural communities.

The objective of this study was to evaluate, through a pilot study, the biophysical suitability of Fingerponds in East African freshwater natural wetlands. The existing wetland uses were studied in order to understand how integrated Fingerpond systems (fisheries and agriculture) can fit into the existing wetland uses. Determinants of aquaculture-agriculture systems functioning, such as site characteristics, water supply, pond stocking, and inputs were observed. This information was used to infer the biophysical suitability of Fingerponds in the natural freshwater wetlands.

Methods

The study was carried out at the Nyangera and Kusa Fingerponds experimental sites at the shores of Lake Victoria, Kenya (Chapter 1).

Natural wetland products and their local uses
Information on wetland products and their uses by the community were obtained through a semi-structured survey. This was aimed at understanding the existing wetland uses vis-à vis the additional use for integrated aquaculture–agriculture (i.e., Fingerponds). Some products harvested from the wetlands were noted by observation of wetland goods transported by the local people passing near the Fingerponds sites.

Soil sampling and analysis
Soil samples were collected at 10-12 cm below the soil surface. The samples were taken to the Soil Science department at Egerton University, Njoro, Kenya for bulk density, particle size and physical parameter analyses. Analysis followed procedures outlined by Okalebo et al. (2002) and Gee and Bauder (1986).

Flood cycle

Information on the annual flood cycle was collected through monitoring of flood event(s) at each site. The dates of flood and the period of flood were observed and recorded for each site during the study period.

Natural fish stocking in the Fingerponds systems

After flood recession, a fish census was carried out in each pond at each site to determine the stocking density and fish stock composition. Fish removal was achieved by seining through each pond at least three times with a 6.5 mm mesh size seine net. Fish counts and identification were done while keeping the catch restrained in a mosquito netting bag set at a corner of the pond. Fish population was estimated using Microfish 3.0 software (Van Deventer and Platts, 1985).

Manure availability and quality

Manure for pond fertilization was supplied by the local community living adjacent to the Fingerponds. Cow manure was the most commonly supplied manure. An estimate of available manure was made indirectly using information on the livestock population from a livestock census conducted by the Ministry of Livestock and Fisheries, Kenya in July 2004 and the approach for estimation of manure production used by Coche et al. (1996, in Aguilar-Manjarez and Nath, 1998). Dried manure samples were collected for nutrient determination in the laboratory.

Results

Common uses of Lake Victoria wetlands in Kenya

Table 2.1 summarizes the most common uses and products of the wetlands in the Kusa area. Most littoral wetlands in East Africa are dominated by papyrus (*Cyperus papyrus*) and act as breeding sites for fish stocks. They also provide a wide range of products and services to the adjacent communities..

Table 2.1: Some common products from Lake Victoria wetlands, Kenya

Product	Main use
Macrophyte biomass	
Papyrus (*Cyperus papyrus*)	Mat making, ropes for house construction, craft making (chairs etc.), thatching material, fuel wood particularly dried rhizomes
Phragmites sp	Thatching, crafts, house construction
Typha sp	Thatching
Cyperus spp.	Thatching and livestock fodder
Fish	Food and income
Clarias gariepinus (local name *Mumi*), *Protopterus aethiopicus* (local name *Kamongo*) and *Shilbe spp.* (local name *Sire*)	
Agricultural crops (seasonal)	Food and income
Water	Domestic use and livestock
Clay	Bricks, hut walls

Table 2.2: Soil characteristics at the Lake Victoria Fingerponds in Kusa and Nyangera, Kenya (values are mean ± standard error except for particle sizes expressed as percentages, n=12)

	SITE	
	Nyangera	Kusa
% clay	59	68
% silt	24.3	21.5
% sand	16.8	10.4
Bulk density (g/cm^3)	1.25 ± 0.05	1.28 ± 0.02
pH	6.78 ± 0.08	6.66 ± 0.03
EC μScm^{-1}	1248 ± 73.5	6215 ± 400
CEC (meq/100g)	41.25 ± 2.3	53.75 ± 4.1

These ecosystems are indeed crucial for the livelihoods of the riparian communities (Mafabi and Taylor, 1993). Among the major products harvested are: plant biomass, fish, seasonal agricultural crops (vegetables, e.g. kales *Brassica oleracea*, tomatoes *Lycoperscicon esculentum*, arrow roots *Dioscorea* sp.). Some of these products constitute the backbone of the rural economies and hence the lifeline for the communities around the lake.

Soil characteristics

Site soil characteristics, particularly particle size distribution, organic matter composition and element composition have critical implications for the performance of any kind of farming system. The soils from the two wetlands are predominantly clay (Table 2.2).

The soil pH was close to neutral in both sites. The electrical conductivity was almost five times higher in Kusa than in Nyangera. The CEC (cation exchange capacity) was also higher in Kusa, probably due to the presence of sodium-alumino-silicates in the soil.

Fingerponds flooding and natural fish stocking

Table 2.3 shows the flood regime at the experimental Fingerponds sites in Kenya. In Kusa, the major flooding period occurred at the beginning of May and lasted for 1-2 weeks. During this period, the water level in the wetlands was 10-15 cm above the ground level. The flood was mainly due to overflow from the river Nyando. Natural fish stocking of Fingerponds occurred during this period. In Nyangera the ponds were submerged for nearly two months in the period May-July.

Table 2.3: Flood cycle in Fingerponds sites in Kenya during the year 2002 and 2003.

Year	2002		2003	
Site	Major flood period	Duration	Major flood period	Duration
Kusa	May	1 week	May	1.5 weeks
Nyangera	May-July	2 months	May-July	2 months

Table 2.4: Fish species composition in the two Kenyan Fingerponds

Fish species	Ecological status in Lake Victoria
Oreochromis niloticus (Linnaeus, 1758)	Introduced
O. variabilis (Boulenger,1906)	Native
O. leucostictus (Trewavas, 1933)	Introduced
Clarias gariepinus (Burchell, 1822)	Native
Aplocheilichthys sp.	Native
Ctenopoma muriei (Boulenger,1906)	Native
Protopterus aethiopicus Heckel, 1851	Native
Haplochromis spp.	Native/endemic

After flood recession a census was carried out in each pond at each of the two study sites. The fish that migrated to the Fingerponds are characteristic of the Lake Victoria fish stocks (Table 2.4). The dominant fish species were three species of tilapia: *Oreochromis niloticus*, *O. variabilis* and *O. leucostictus*.

The study revealed that Fingerponds can be stocked adequately by natural floods. The natural stocking densities averaged 11 and 3 fish m^{-2} in Nyangera and Kusa, respectively. The duration of the Fingerponds season may vary from one location to another. After the normal annual flooding period, the ponds may retain adequate water for fish culture for at least six months before the water level declines to a critical depth (Table 2.5). However, there could be short seasons after unexpected flooding during the short rains as was observed between the end of December 2002 and March 2003.

Animal manure availability for integrated wetlands aquaculture-agriculture
The use of farmyard manure and agricultural by-products to improve soil fertility and boost crop production is a common phenomenon in many rural communities in Africa. If livestock production is known, available manure for pond fertilization can be estimated. However, due to lack of such data, the livestock numbers in a specific area was used as a surrogate measure of manure availability. Table 2.6 shows the livestock population density and estimated manure production in Kayano village (Kusa) in 2004.

In Kenya, dry cow manure was used to enrich the ponds at an application rate of 1.25-2.5 tonnes per hectare per two weeks. Manure was supplied by the villages adjacent to the ponds. The quality of manure used at the two Fingerponds experimental sites are shown in Table 2.7.

Table 2.5: Duration of Fingerponds seasons in experimental sites at the shores of Lake Victoria, Kenya

Year	Duration of Fingerpond season (months)	
	Nyangera	Kusa
2002	6	6
2002/2003*	3	2.5
2003/2004	8	6

*There was an unexpected flood at both sites in December 2002 followed by a short Fingerponds season

Table 2.6: Livestock population and estimated manure production in Kayano village, Kusa, Kenya (unpublished livestock census data; GOK, Ministry of Livestock and Fisheries)

Animal type	Number of livestock	Total DM wt tonnes /day	Estimated manure production (tonnes/year)
Cattle	234	0.737	265
Goat	129	0.068	24
Sheep	102	0.054	19
Poultry	181	-	-
Total			308

Table 2.7: Manure quality at the Kenyan Fingerpond sites (values are mean values of total nitrogen and total phosphorus in manure samples ± standard error, n=6)

Site	Manure type	Manure quality	
		TN mg/g	TP mg/g
Kusa	Dry cattle manure	18.6 ± 1.45	3.21 ± 0.06
Nyangera	Dry cattle manure	19.67 ± 1.2	3.29 ± 0.19

Fingerpond products

The main aim of Fingerponds is to enhance the production of fish and seasonal agricultural products for the poor rural communities living at the edge of freshwater in East Africa. The fish yields ranged from 402-1068 kg/ha in manured ponds and 180-423 kg/ha in un-manured ponds (Figure 2.2). The production of local vegetables *Brassica oleracea* or *sukuma wiki* averaged 17 tonnes in Nyangera site per hectare per year .

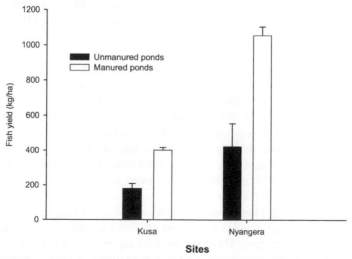

Figure 2.2: Fish yields in a polyculture experimental Fingerponds in Kenya (n=3 harvests and average culture period is 5-6 months for Kusa and Nyangera sites respectively)

Discussion

Natural wetlands as lifeline for riparian communities

Natural wetlands constitute an important resource for the local communities living around them. In Kusa, for instance, papyrus biomass harvesting, and particularly culms for mat making is one of the most important sources of livelihood for the community. Harvesting is done throughout the year and products are sold locally on weekly market days. However, the price of papyrus slumps whenever other resources decline (especially due to drought) and the whole community resorts to papyrus harvesting, thus flooding the market (J. Kipkemboi, personal observation). This exposes the local people to economic hardship and hence there is a need to diversify the sources of livelihood.

The basis of Fingerponds as an intermediate technology is built on existing uses of these ecosystems by the local people particularly seasonal agriculture and fishery. Since the local communities are already aware of some of these wetland benefits, such technology may be adopted easily. However participatory research is necessary to ensure that appropriate technology is adapted to specific sites.

Wetland annual flooding and Fingerponds fish stocking

Lake Victoria freshwater wetlands are characterized by annual water fluctuation. During heavy rain seasons in the catchment, the water volume in the lake increases and some areas around the shoreline are inundated. Flood extent depends on the amount of precipitation received in the catchment and may vary in extent and duration from year to year and between different sites of the lacustrine wetland. For example, the Nyangera site floods for about two months while in Kusa the flood lasts for about a week only. During this time fish move within the aquatic macrophyte vegetation zone and may breed in the wetland. In the case of the Rufiji floodplain in Tanzania, floods may last for a whole year or may not come at all (H. Lamtane, pers. com) while in Ugandan Fingerpond sites at Lake Victoria, the floods were insufficient to stock the ponds. This introduces an element of risk for this technology.

During the flood period, fish migrate to the ponds through the wetland vegetation and colonize the Fingerponds. The fish species observed in the Kenyan sites are typical of Lake Victoria. Notably absent in the Fingerponds was the Nile perch, *Lates niloticus*, which is not able to migrate through the wetland due to low tolerance to hypoxia (Chapman et al., 2002). Ochumba and Manyala (1992) indicated that Lake Victoria fish can swim several kilometers upstream on the Sondu Miriu River. This may explain why the fish in the Kusa Fingerponds, located in the floodplain of river Nyando and about 5 km from Lake Victoria, contained fish species commonly found in the lake.

The unique feature of Fingerponds is their capability of self stocking. The fish stock is a polyculture that consists of several species capable of migrating through the natural wetland vegetation. This kind of stocking may be advantageous in that different fish utilize various ecological niches and a variety of food chains within the pond and hence improve pond productivity. Undesirable fish species and sizes may be removed as a way of biological manipulation of the systems in order to enhance productivity.

Suitability of Fingerponds to East African wetlands; lessons learned from experimental Fingerponds, Kenya

Seasonal wetland fishery and vegetable production are common practices in many parts of Lake Victoria littoral wetlands. The principle of Fingerpond technology rekindles the existing traditional knowledge of post-flood wetland fishery among the local communities. This makes it an attractive venture as the concept is not alien to the users. When the Fingerponds concept was first discussed with the local community, they had no doubts that this technology would work under local conditions. Fingerponds can be adequately stocked by floods. This reduces the cost of purchasing fingerlings and associated logistics of transportation which may be one of the impediments to conventional aquaculture. The soils in most wetlands, which are predominantly clay, are suitable for aquaculture. Dry season vegetable production practiced by the local communities at the landward edge of the wetlands can be incorporated into Fingerpond systems. However in some wetlands, there may be a few spots with sodic soils which may limit crop production (Mati and Mutunga, 2003). This study did not investigate the effect of high electrical conductivity in pond water caused by soil salinity on pond phytoplankton and zooplankton dynamics. There is a need for a study on how this affects fish growth.

The flood regime and amplitude may vary from year to year (Kipkemboi, personal observation). This implies that Fingerponds may either remain under water for a longer time than expected, or may not be flooded at all but can be filled with groundwater seepage or rainwater. If the latter happens, it is possible to stock Fingerponds using fish from other sources such as Fingerponds in other sites of from the lake or river.

Enhanced benefits from wetlands

The products from Fingerponds comprise of fish and vegetables. Natural wetland fishery has been declining over the past decade (Balirwa, 1998). Fingerponds may provide an opportunity for enhancing food production from natural wetlands. This study revealed that the net fish yields were higher in manured ponds compared to unmanured ones, although fish yields from manured Fingerponds were lower than those reported in conventional systems (Lin et al., 1997). Fingerponds may supplement protein and vitamins to the adjacent villages through additional fish and vegetable supply. As a result this may alleviate malnutrition problems especially among the children. Fingerponds products are not intended primarily for trade; however surplus may be sold for income. Increased productivity may also reduce the encroachment on the wetlands for extensive cultivation currently observed in many developing countries.

Conclusion

Communities living along Lake Victoria wetlands derive a number of products from natural wetlands for their livelihoods, mainly in the form of emergent macrophyte biomass, fishery and seasonal agriculture. To enhance the declining wetland resources, smallholder seasonal agriculture can be practiced alongside pond aquaculture. Fingerponds will augment the existing wetland benefits through fish and vegetable production. If sound scientific principles are integrated with the existing indigenous knowledge, sustainable use of natural wetlands is likely to be achieved. The soils in natural wetlands around Lake Victoria in Kenya are generally suitable for aquaculture and if well drained can support crop production. Annual wetland flooding has a dual role in Fingerponds in that it provides fish stocks and acts as water supply for these systems. Manure supplements to improve pond productivity can be supplied adequately from the adjacent villages as long as livestock farming is practiced by the local community.

Although most of the lacustrine and floodplain wetlands are suitable for Fingerponds, careful site selection is important for the success of these systems. There is a need for more research on socio-economic aspects and general sustainability of such systems.

Acknowledgements

I wish to acknowledge the financial support from the European Union Fingerponds project contract no. ICA4-CT-2001-10037. Additional funding for fieldwork was provided by the International Foundation for Science (IFS) through grant no W/3427-1. I also wish to thank Dr. Johan Rockström for his comments on the initial drafts of this chapter and assistance at the beginning of the project, particularly in establishing contacts with local institutions such as the Relma-Sida project.

References

Aguilar-Manjarez, J., Nath, S.S., 1998. A strategic reassessment of fish potential in Africa. CIFA Technical paper No. 32, Rome, FAO, 170 pp.

Balirwa, J.S., 1998. Lake Victoria wetlands and the ecology of the Nile tilapia, *Oreochromis niloticus* Linné. Ph.D. Thesis, Agricultural University of Wageningen and IHE-Delft, The Netherlands. A.A. Balkema Publishers, Rotterdam,. 247 pp.

Brummett, R.E., Williams, M.J., 2000. The evolution of aquaculture in African rural and economic development. Ecological Economics 33,193-203.

Chapman, L.J., Chapman, C.A., Nordlie, F.G., Rosenberger, A.E., 2002. Physiological refugia; swamps, hypoxia tolerance and maintenance of fish diversity in Lake Victoria region. Comparative Biochemistry and Physiology Part A 133, 421-437.

Coche, A.G., Muir, J.F., Laughlin, T., 1996. Management for freshwater fish culture. Ponds and water practices. FAO Training Series, 21/1, 233 pp.

COFAD, 2002. Back to basics: Traditional inland fisheries management and enhancement systems in sub-Saharan Africa and their potential for development. Deutche Gesselschaft fur Technische Zussamenarbeit (GTZ) GmbH, Eschborn, 203 pp.

Denny, P., 1985. The structure and function of African euhydrophyte communities. In: P. Denny (ed.), The ecology and management of African wetland vegetation. Dr. W. Junk Publishers, Dordrecht, pp. 125-151.

Denny, P., 1989. Wetlands. In: Strategic resources planning in Uganda. UNEP Report IX, 103 pp.

Denny, P., Kipkemboi, J., Kaggwa, R., Lamtane, H., 2006. The potential of Fingerpond systems to increase food production from wetlands in Africa. International Journal of Ecology and Environmental Sciences, 32(1), 41-47.

Devendra, C., 1995. Integration of agriculture and fish farming in Indonesia. In: J.-J. Symoens, J.-C. Micha (eds.), Seminar "The management of integrated freshwater agro-piscicultural ecosystems in tropical areas", Brussels, 16-19, May 1994,

Technical Centre for Agricultural and Rural Co-operation (CTA) Royal Academy of Overseas Sciences, Brussels, pp. 329-341.

Ellenbroek, G.A., 1987. Ecology and productivity of an African wetland system, The Kafue Flats, Zambia. Dr. W. Junk Publishers, Dordrecht, 267 pp.

FAO, 2000. Small ponds make a big difference: Integrating fish with crop and livestock farming. http://www.fao.org/docrep.

Fernando, C.H., Halwart, M., 2000. Possibilities for the integration of fish farming into irrigation systems. Fisheries Management and Ecology 7, 45-54

Gee, G.W., Bauder, J.W., 1986. Particle size analysis. In: A.Klute (ed.) Methods of soil analysis; Part 1 Physical and mineralogical methods. 2nd edition, American Society of Agronomy, Madison, Wisconsin USA, pp. 383-408.

Haylor, G., Bhutta, M.S. 1997. The role of aquaculture in the sustainable development of irrigated farming systems in Punjab, Pakistan. Aquaculture Research 28, 691-705.

Kangmin, L., Pehzein, L., 1995. Integration of agriculture, livestock and fish farming in the Wuxi region of China. In: J.-J. Symoens, J.-C. Micha (eds.), Seminar "The management of integrated freshwater agro-piscicultural ecosystems in tropical areas", Brussels, 16-19, May 1994, Technical Centre for Agricultural and Rural Co-operation (CTA) Royal Academy of Overseas Sciences, Brussels, pp. 309-328.

Kapetsky, J.M., 1994. A strategic assessment of warm-water fish farming potential in Africa. *CIFA Technical Paper*, No. 27. Rome, FAO, 67 pp.

Kapetsky, J.M., 1995. A first look at the potential contribution of warm water fish farming to food security in Africa. In: J.-J. Symoens, J.-C. Micha (ed.), Seminar "The management of integrated freshwater agro-piscicultural ecosystems in tropical areas", Brussels, 16-19, May 1994, Technical Centre for Agricultural and Rural Co-operation (CTA) Royal Academy of Overseas Sciences, Brussels, pp. 547-572.

Lin, C.K., Teichert-Coddington, D.R., Green, B.W., Veverica, K.L., 1997. Fertilization regimes. In: Egna, H.S., Boyd, C.E. (ed.), Dynamics of Pond Aquaculture, CRC Press, Boca Raton, Florida, pp.73-107.

Mafabi, P., Taylor, A.R.D., 1993. The national wetlands programme, Uganda. In Davis, T.J. (eds.) Towards wise use of wetlands, Wise Use Project, Ramsar Convention Bureau, Gland, Switzerland, pp 52-63.

Mati, B.M., Mutunga, K., 2003. Integrated soil fertility and management assessment of the Kusa profile, Lake Victoria basin. Technical report for Regional Land management Unit (Relma) Kusa pilot project, 89 pp.

Ochumba, P.B.O., Manyala, J.O., 1992. Distribution of fishes along the Sondu-Miriu River of Lake Victoria, Kenya with special reference to upstream migration, biology and yield. Aquaculture and Fisheries Management, Oxford Vol.23, No 6, pp 701-719.

Okalebo, J.R., Gathua, K.W., Woomer, P.L., 2002. Laboratory methods of soil and plant analysis; A working manual. 2nd edition, SACRED Africa, Nairobi, 128 pp.

Prein, M., 2002. Integration of aquaculture into crop-animal systems in Asia. Agricultural Systems 71, 127-146.

Pullin, R.S.V, Prein, M., 1995. Fishponds facilitate natural resources management on small-scale farms in tropical developing countries. In: J.-J. Symoens, J.-C. Micha (eds.), Seminar "The management of integrated freshwater agro-piscicultural ecosystems in tropical areas", Brussels, 16-19, May 1994, Technical Centre for Agricultural and Rural Co-operation (CTA) Royal Academy of Overseas Sciences, Brussels, pp. 169-186.

Silvius, M.J., Oneka, M., Verhagen, A., 2000. Wetlands: lifeline for people at the edge. Phys. Chem. Earth (B) 25 (7-8), 645-652.

Symoens, J.-J. 1995. Sustainable use of wetlands: respect for environment and biodiversity. In: J.-J. Symoens, J.-C. Micha (eds.), Seminar "The management of integrated freshwater agro-piscicultural ecosystems in tropical areas", Brussels, 16-19, May 1994, Technical Centre for Agricultural and Rural Co-operation (CTA) Royal Academy of Overseas Sciences, Brussels, pp. 87-107.

Van Deventer, J.S. and Platts, W.S., 1985. A computer software system for entering, managing and analyzing fish capture data from streams. USDA Forest Service Research Note INT-352, Intermountain Research Station, Ogden, Utah, 12 pp.

Vincke, M.M.J., 1995. The present state of development in continental aquaculture in Africa. In: J.-J. Symoens, J.-C. Micha (eds.), Seminar "The management of integrated freshwater agro-piscicultural ecosystems in tropical areas", Brussels, 16-19, May 1994, Technical Centre for Agricultural and Rural Co-operation (CTA) Royal Academy of Overseas Sciences, Brussels, pp. 27-61.

Chapter

3

Effects of soil characteristics and hydrology on the functioning of smallholder wetland aquaculture-agriculture systems (Fingerponds) at the shores of Lake Victoria, Kenya

Abstract

Experimental smallholder wetland-based integrated aquaculture-agriculture systems called 'Fingerponds' were established at two sites (Nyangera and Kusa) at the shores of Lake Victoria in Kenya. The overall aim was to enhance the wetland fishery potential. The soil textural class was clay in both sites and was generally suitable for pond aquaculture. Soil electrical conductivity and sodium levels were significantly higher in Kusa compared to Nyangera (t-test, $P<0.001$ and $P<0.05$, respectively). The presence of patches of soils with encrustations of sodium salts at the wetland margin in Kusa affected the overall functioning of these systems and emphasizes the need for careful site selection. In Fingerponds, the water supply is un-regulated and the water balance is maintained by natural losses and gains. At the beginning of the season, flood events are critically important for the initial water supply. During the functional period of the ponds (which lasted for about 6 months after flood recession), precipitation accounted for nearly 90% of the total water gains whilst seepage and evaporation contributed an average of 30 to 70% of the losses, respectively. Seasonal pond water budgets indicated that the losses outweighed the gains leading to a progressive decline of water depth during the dry season. A prediction of the effect of pond volume and weather conditions on the functional period was carried out using a dynamic simulation model. The results indicated that the culture period can be extended by 2½ months by deepening the ponds to an average depth of 1.5 m. Drier weather accelerated losses and shortened the culture period by 1-2 months.

Key words: Lake Victoria wetlands, Kenya, water balance, integrated aquaculture systems, Fingerponds, STELLA.

Publication based on this chapter:
Kipkemboi, J., van Dam, A.A., Mathooko, J.M., Denny, P. Effects of soil characteristics and hydrology on the functioning of smallholder wetland aquaculture-agriculture systems (Fingerponds) at the shores of Lake Victoria, Kenya. Aquacultural Engineering (submitted).

Introduction

Floodplains and littoral wetlands have potential for the enhancement of small-scale fish production for rural communities (Welcomme, 1975, Welcomme and Bartley, 1998). In spite of this, limited effort has been made to explore the possibilities of sustainable exploitation of this resource particularly in East Africa. The potential of wetlands for augmenting terrestrial agricultural productivity is increasingly recognized (FAO, 1998; McCartney et al., 2005) but can lead to degradation of the wetland. However, seasonal agriculture as commonly practiced in many African wetlands combined with small-scale aquaculture to enhance food production is feasible and potentially sustainable.

The potential for aquaculture for Africa was demonstrated by Kapetsky (1994) and Aguillar-Manjarez and Nath (1998). The challenge of making this a reality remains unresolved amidst increasing poverty in the region. Integrated aquaculture and farming in wetlands is one way of increasing protein production and food security in seasonally flooded wetlands (Fernando and Halwart, 2000; Halwart and van Dam, 2006). Smallholder integrated fish and vegetable production systems (called "Fingerponds" because they appear like fingers into the emergent macrophyte zone from a birds eye view) were trialed in natural wetlands around Lake Victoria, Kenya. The idea originates from existing traditional floodplain/littoral wetland fisheries in Africa and adaptations of Asian integrated aquaculture-agriculture systems (Denny, 1989; Denny and Turyatunga, 1992). Earthen ponds are excavated in fringe wetlands during the dry season and the soil is spread beside the depressions to create raised beds for vegetable production. The ponds resemble natural flood pools used for wetland fish capture by local communities while the gardens are a continuation of normally-existing seasonal swamp margin vegetable patches. Indeed, Fingerponds are an extension of some of the wetland's existing fishery and agriculture functions. The ponds are stocked with wild fish during annual flooding of the wetland. Fish culture and garden management start after flood recession. Livestock manure and vegetable wastes are added to the ponds to stimulate the food-chain for fish while water from the ponds may be used for irrigation (Denny et al., 2006).

While Fingerpond technology is promising, particularly for poor rural riparian communities, an understanding of its biophysical functioning is the starting point of its implementation and management. In aquaculture, soil properties and topography play significant roles not only in siting of the ponds but also in their overall functioning (Boyd et al., 1999). Very often, the water balance has been ignored in pond aquaculture studies in sub-Saharan Africa although a good understanding of the water budget is essential for the management of these systems (Boyd, 1982; Nath and Bolte, 1998; Boyd and Gross, 2000; Braaten and Flaherty, 2000). The objectives of this study were to: (1) assess the soil characteristics at the two experimental Fingerponds sites; (2) to evaluate the effects of the flooding regime on pond water supply and assess the hydrological variables that determine the water balance and consequently the pond functional period (defined below) after flood recession.

Methods

Site description

Topographic and climatic characteristics

Topography is an important feature in the excavation of ponds (Kelly and Kohler, 1997). The slope of the site was 1.1% in Nyangera and 0.33% in Kusa. Flat areas are suitable for Fingerponds since they can flood easily when water levels rise in the lake or river. Most areas around Lake Victoria on the Kenyan side fall within the dry semi-arid humid agro-ecological zone (FAO, 1996) characterized by low to moderate suitability for rain-fed agriculture. Rainfall is erratic and may fall in high intensities. However, since evapotranspiration is high, loss of soil moisture is also high leading to intra-seasonal dry spells (Kipkemboi, personal observation.). The average daily solar radiation recorded in the Kusa weather station between May 2003 and February 2004 was 230.02 ± 38.81 watts m^{-2} and the mean temperature between 0900 and 1500 hours fluctuated between 23.6 and 27.8 $^{\circ}$C. Temperatures drop at night with the lowest values just before sunrise and high peaks occurring late in the afternoon (Figure 3.1). The relative humidity in Kusa fluctuated around 70% but decreased to less than 40% during the dry season (Figure 3.2).

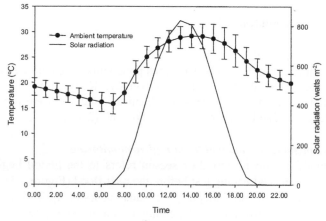

Figure 3.1: Mean solar radiation (Watts m^{-2}) and ambient temperature ($^{\circ}$C) variation during the day in Kusa, Kenya (May 2004-March 2005)

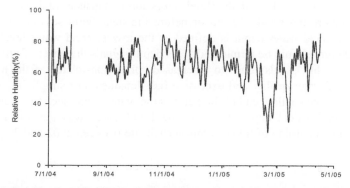

Figure 3.2: Relative humidity (%) variation in Kusa, Kenya (the gap during August was due to technical problems with the weather station).

Fingerpond design and construction
The ponds are rectangular, measuring 24 m × 8 m with a depth of 1 m at the shallow end to 2 m at the deep end. At each site, four adjacent ponds and gardens were constructed manually by the local communities. Due to variability in the geological characteristics between sites and limitations associated with groundwater intrusion during construction, the actual pond depths and volumes varied slightly from the initial design. Table 3.1 shows the layout and the characteristics of Fingerponds.

Table 3.1: Design parameters and variability in actual characteristics of Fingerponds in the two study sites in Kenya (Mean ± SD, n=4)

Attribute	Initial design	Nyangera		Kusa	
Pond area (m^2)	192	187.53	± 3.33	190.92	± 2.53
Maximum pond depth (m)	2	1.31	± 0.9	1.44	± 1.8
Minimum pond depth (m)	1	0.82	± 0.35	0.96	± 0.34
Average depth (m)	1.5	0.93	± 0.84	1.04	± 0.25
Pond volume (m^3)	242	168.33	± 7.05	194.4	± 24.20
Garden area (m^2)	192	≈192		≈192	
Estimated water shed area (m^2)		≈48		≈48	

Data collection and analysis

Soil sampling and analysis
Composite samples from the newly excavated pond soils were collected and taken to the Soil Science Department, Egerton University, Kenya for physico-chemical characterization. Soil particle size and chemical analysis followed procedures outlined by Gee and Bauder (1986) and Okalebo et al. (2002).

Fingerponds functional period and flood regime monitoring
The Fingerponds functional period or season refers to the period beginning from flooding of the ponds and ending just before the next flood (Figure 3.3). Flooding in the Lake Victoria littoral and floodplain wetlands occurs in April/May and may last between two weeks to two months, depending on the proximity to the lake and the size of flood. The season ends in April, just before the next flooding period. The functional period with respect to pond aquaculture starts after flood recession, disconnection of the ponds from the floodwaters and fish census; and ends in the dry season just before the ponds dry up or before the next flood season - whichever comes first. Critical water depth is defined as the lowest pond water level beyond which fish culture should be terminated. This depth signals the onset of unfavorable conditions in the ponds due to increased predation-related fish mortality, deterioration of water quality and excessive fish densities. For simplicity, exposure of the bottom at the shallow end of the pond as the water depth declines is used as an indicator of the critical depth. Data on the flood cycle was generated through observation and recording of flood events at each site between 2002 and 2005.

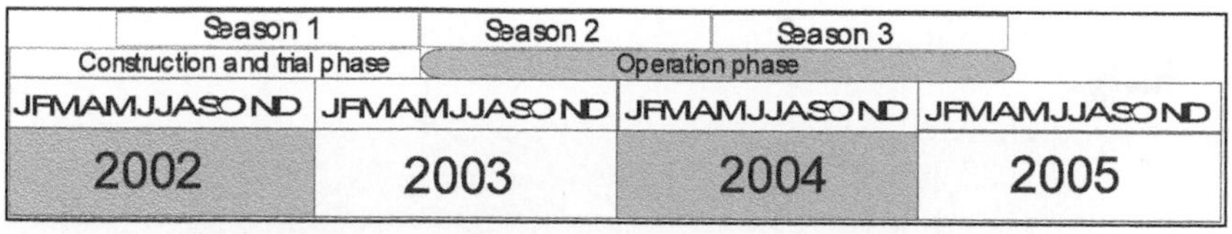

	Season 1	Season 2	Season 3	
Construction and trial phase		Operation phase		
JFMAMJJASOND	JFMAMJJASOND	JFMAMJJASOND	JFMAMJJASOND
2002	2003	2004	2005

Figure 3.3: Definition of events and Fingerponds seasons during the experimental period. From top to bottom: Fingerponds seasons, site activities, months and calender years

Hydrological measurements

Hydrological data were collected from weather stations set up at each site. The stations consisted of a Class A evaporation pan, a thermometer and a standard rain gauge. An automatic field weather station (Weather Hawk 240, Campbell Scientific, USA) was installed in Kusa to augment hydrological data collection.

Hydrological measurements were carried out daily between 0800 and 0900 hours. Pond water level changes were monitored with staff gauges installed permanently at the middle of each pond to estimate average water depth. Pond evaporation was estimated from an evaporation pan and corrected using a pan coefficient of 0.81 (Green and Boyd, 1995). Precipitation depth was measured with a rain gauge. Groundwater level fluctuations were monitored daily with piezometers installed along a transect through the pond area using a measuring tape and sounding device. Ambient air temperature was recorded twice daily at each site between 0800-0900 hrs and 1400-1500 hrs. Additional data collected from the automatic weather station from May 2004 to April 2005 included relative humidity, solar radiation (Watts m^{-2}), wind speed (m/sec), and air temperature.

Data analysis

Statistical analysis   Statistical data analysis was performed using SPSS 11.0 (SPSS Inc., Chicago). T-tests were employed to compare the soil parameters between the study sites while correlations were used to quantify the relationship between the pond water and groundwater level variation. Data were reported as mean ± standard deviations (SD). Means were declared significantly different at alpha levels of 0.05, unless stated otherwise.

Hydrological calculations   The pond water budget was calculated from the hydrological equation:

$$Gains = Losses \pm Change\ in\ storage \qquad (3.1)$$

Fingerponds are unique in that they lack regulated flows (Figure 3.4). The possible water gains are: initial filling by flood (F), precipitation (P), runoff (R) and seepage into the ponds (S_i) while the losses include evaporation (E), seepage out (S_o) and possibly abstraction (A) for irrigation of the raised-bed gardens. The watershed area contributing to pond water gains through runoff was assumed to consist mainly of the raised-bed garden. The hydrological equation for the ponds is therefore:

$$F + P + R + S_i = E + S_o + A \pm \Delta V \qquad (3.2)$$

where ΔV is the change in volume indicated by staff gauge water depth changes (ΔH).

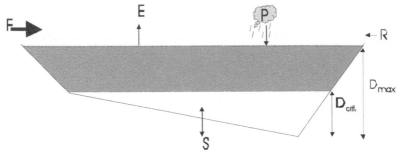

Figure 3.4: Schematic drawing of a Fingerpond showing water gains and losses (D_{max} is the maximum depth after flood recession while D_{crit} is the lowest water depth which signals harvesting of the ponds)

For convenience, hydrological measurements of volume are expressed as depths in millimeters or centimeters. The initial flood was ignored in the water budget estimation since it only occurs once at the beginning of the season and fills the pond to the maximum level. The equation for Fingerponds water budget is therefore expressed as:

$$P + R + S_i = E + S_o + A \pm \Delta V \tag{3.3}$$

Seepage is often a difficult hydrological variable to determine accurately. In earthen ponds, particularly those located in areas where the groundwater level exhibits seasonal fluctuations, seepage may occur in both directions. It is not easy to separate seepage loss from seepage gains; however it is possible to determine the direction and the rate of net seepage. Daily net seepage rate (S_n) was estimated through the process of elimination from Equation 3 (Boyd and Gross, 2000). This approach is only applicable during dry weather, when there is no precipitation and runoff (P=R=0). In this approach, the changes in staff gauge depth readings between consecutive days in a rainless period were obtained and corrected for evaporative water loss:

$$S_n = \Delta H - E \tag{3.4}$$

where ΔH is the water level change during the period under consideration. The net seepage was expressed as rate per unit time (mm day^{-1}).

Runoff was estimated using the approach developed by the U.S. Soil Conservation Service (SCS) (1972). In this method, a combination of soil texture; hydrological grouping and land use/vegetation cover determined through visual observation is used to assign a runoff curve number. Using the daily precipitation data measured at the sites, an estimated runoff depth was computed as:

$$R = \frac{(P_d - 0.2S)^2}{P_d + 0.8S} \tag{3.5}$$

where R is the runoff depth (mm day^{-1}), P_d is the daily precipitation measured on site and S is the maximum watershed retention (mm day^{-1}) computed as follows:

$$S = \left(\frac{1000}{CN} - 10 \right) \times 25.4 \qquad (3.6)$$

CN is the curve number for the combination of soil type, land use and hydrological condition and was obtained from the SCS table. A runoff curve number (CN) of 91, corresponding to soil group D (clay soils with low infiltration) and cultivated area with row crops was used to estimate the runoff depth.

Pond water level dynamic simulation models A dynamic hydrological model was developed using STELLA 8.0 (High Performance Systems, Lebanon, USA) to simulate the pond water depth variation and estimate the potential duration of the functional period of Fingerponds under different scenarios. The model was based on similar approaches used for modelling hydrological processes in wetlands by Spieles and Mitsch (2000) and Zhang and Mitsch (2005). Simulations used Euler integration and a time step of one day. A conceptual STELLA model diagram with the pond water volume (expressed as depth) as the state variable and precipitation, runoff, evaporation and seepage as rate variables is shown in Figure 3.5.

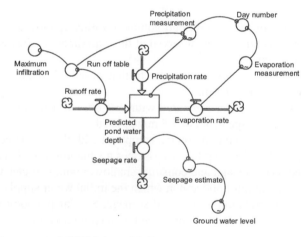

Figure 3.5: A conceptual STELLA model diagram

Model Equations:

Pond_water_depth(t) = Pond_water_depth(t - dt) + (Precipitation_rate + Runoff_rate - Evaporation_rate - Seepage_rate) * dt
INIT Pond_water_depth = 1040
INFLOWS:
Precipitation_rate = Precipitation_table
Runoff_rate = Run_off_table
OUTFLOWS:
Evaporation_rate = if Pond_water_depth>0 then Evaporation_measurement else 0
Seepage_rate = Seepage_table
Day_number = INT(TIME)+1
Run_off_table = (Precipitation_table-(0.2*25.1)^2)/(Precipitatation_table+0.8*25.1)

Using the daily water balance parameters, this approach allows prediction of the functional period under different scenarios. The description of the model variables is presented in Table 3.2. Two factors that may play an important role in determining the length of Fingerponds season are; the maximum capacity of the pond, and the hydrological conditions after flood recession. Pond water level variation was simulated based on the following two scenarios.

Table 3.2: Model variables

Variable	Units	Source
Precipitation (P)	mm day^{-1}	Field measurements
Runoff (R)	mm day^{-1}	calculated
Evaporation (E)	mm day^{-1}	Field measurements
Seepage (S)	mm day^{-1}	Calculated and calibrated with Boyd and Gross (2000).
Groundwater level	mm	Field measurements
Pond volume[1]	mm	simulation

[1]For convenience volume is expressed as depth (mm)

Scenario A: The effect of maximum pond volume on Fingerponds season duration under normal weather conditions as observed at the experimental sites in Season 2. Three situations were simulated.
1) A pond with a maximum volume similar to that observed in the experimental ponds, with an average depth of 1 m.
2) A pond with a larger volume compared with that observed in the experimental ponds. For this case an average depth of 1.5 m was used.
3) A pond with a smaller maximum volume compared with the experimental ponds, i.e. an average depth of 0.75 m depth. This assumes a situation in which the farmers do not adhere to the layout design and construct shallower ponds. It can also be used to evaluate a situation without flood and in which the initial water supply is achieved only through direct precipitation, runoff and seepage. Similar phenomena occurred in Season 3 when the ponds filled to about ¾ of the maximum volume.

Scenario B: The effect of weather conditions on water supply after flood recession (considering a pond volume of scenario 1b above) on the Fingerponds season. The model simulation was based on responses to three possible situations.
1) Normal weather based on observed hydrological conditions at the experimental sites in Season 2.
2) Dry weather conditions after flood recession, with ¼ less precipitation and increased evaporation by a similar magnitude.
3) Drier weather conditions with rapid decline in groundwater and accelerated seepage.

Results

Soil characteristics

The soil textural class was clay in both study sites (Table 3.3). However, there were significant differences in particle size distribution between the sites; clay (t-test, P <0.001) and for silt and sand (t-test, P<0.05). In Kusa, up to 68% clay was recorded compared to about 59% in Nyangera. The bulk densities were similar in both sites and ranged from 1.16-1.32 g cm^{-3}. Soil pH values were circum neutral. Electrical conductivity was high and values of over 6 mS cm^{-1} were measured in Kusa; significantly higher than in the Nyangera site (t-test, P<0.001). High soil CEC of over 40 meq/100 g, often associated with clay soils, was observed in both sites. Sodium concentration was significantly higher in Kusa than in Nyangera. Both sites were characterized by low to moderate levels of calcium and magnesium in the soils. The concentrations differed between sites for calcium and magnesium, respectively (P<0.05). Total soil nitrogen and phosphorus concentration were generally low and did not differ significantly between sites.

Table 3.3: Summary of pond soil physical and chemical parameters, n=12 and n=4 respectively (values reported as means ± SD)

	Sites				t-statistic
	Nyangera		Kusa		
clay (%)	58.97	± 3.52	68.21	± 1.46	8.40 **
silt (%)	24.32	± 3.408	21.44	± 1.14	2.77*
sand (%)	16.78	± 6.56	10.42	± 0.78	3.34*
Organic matter (%)	8.26	± 1.04	6.74	± 1.01	
Bulk density (g/cm^3)	1.25	± 0.09	1.28	± 0.04	
pH range	6.55	− 6.93	6.69	− 6.71	
EC (mS/cm)	1.23	± 0.15	6.23	± 0.80	12.19**
CEC (meq/100g)	41.25	± 2.36	53.75	± 4.13	2.63*
K (meq/100g)	1.22	± 0.41	1.27	± 1.38	0.21
Na (meq/100g)	1.70	± 3.20	3.20	± 0.53	4.39*
Ca (meq/100g)	2.47	± 0.52	1.06	± 0.43	4.16*
Mg (meq/100g)	1.71		0.95	± 0.14	10.99**
N (ppm)	69.52	± 43.02	80.15	± 16.13	0.46
P (ppm)	26.25	± 11.58	23.5	± 7.50	0.40

Significant difference of mean values for 2-tailed independent samples t-test assuming equal variances at *P<0.05 ** P<0.001

Water supply

Flood regime

Table 3.4 shows the flood regime at the two study sites between 2002 and 2004. In Kusa, the flood lasted for about one week while in Nyangera it lasted for about two months. The flood at both sites coincided with the long rains in the catchment, which normally start in March/April and resulted in a rise in the water level of Lake Victoria in April/May. On average, the water depth rose to about 20 cm above the ground surface in both sites.

Table 3.4: Flood regime at Fingerponds sites during the year 2002 and 2004.

	2002		2003		2004
Site	Flood period	Duration	Flood period	Duration	Flood period
Kusa	May	1 week	May	1.5 weeks	No flood
Nyangera	May-July	2 months	May-July	2 months	No flood

Pond water level dynamics
There was a progressive decline in pond water levels during the dry season. A high variability was observed between seasons and sites: for instance, in Season 2 the Nyangera ponds maintained adequate levels for fish culture for most of the year while in Kusa water levels started to decline below the critical level in November and ponds dried out completely by the end of January (Figures 3.6a and b).

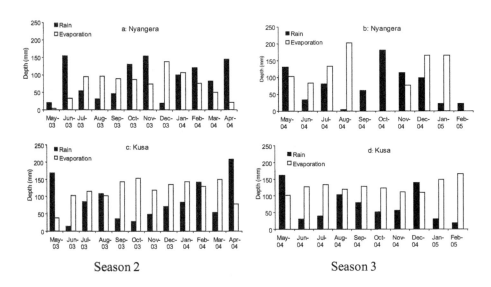

Season 2   Season 3

Figure 3.6: Rainfall and evaporation in Nyangera and Kusa study sites in 2003-2004 and 2004-2005 Fingerponds season (no records of evaporation were available for Nyangera in September and October 2004)

In Season 3, neither site flooded in May. The highest average water level attained at the beginning of the season, due to direct precipitation, groundwater seepage into the ponds as well as runoff was about 70 cm and 80 cm in Nyangera and Kusa, respectively (Figures 3.6 c and d). During the season, the water level declined faster compared to the previous year, particularly in Nyangera, leading to drying of the ponds at the beginning of September. The rapid groundwater level decline during Season 3 explains the variations in pond water level as there was a strong correlation between pond water level and groundwater level variation $r_s(62)= 0.89$ and $r_s(62) = 0.67$ (with $P < 0.01$) for Nyangera and Kusa, respectively.

Hydrological variables and pond water balance

During the second Fingerponds season, rainfall totalled 1056.5 and 1046.1 mm in Nyangera and Kusa, respectively. In Season 3 the rainfall, although measured only for 10 months, from May 2004 and February 2005 was considerably lower and totalled 746.8 mm and 709.2 mm in Nyangera and Kusa, respectively. Rainfall distribution at both study sites did not show a distinct pattern, however, a higher precipitation was recorded in the April-June period (Figures 3.7 a,b,c and d). In Nyangera there was a slight peak in October and November following the short rains. The total annual evaporation varied between the two study sites with 820.3 mm in Nyangera compared to 1409.2 mm in Kusa for Season 2.

Table 3.5 shows the summary of water gains and losses during the two operational Fingerponds seasons from 2003-2005. The total estimated surface runoff during the Fingerpond functional period was 52.1 mm and 26.7 mm in Season 2, and 24.9 mm and 31.0 mm in Season 3 in Nyangera and Kusa, respectively. The mean daily seepage rates varied between sites and seasons. The seepage rate was higher in Nyangera than in Kusa and varied between 2.3 ± 1.8 mm day^{-1} to 4.9 ± 2.7 mm day^{-1} and 1.8 ± 1.6 mm day^{-1} and 2.2 ± 0.9 mm day^{-1} for the two sites in Season 2 and 3, respectively. During the study period, the losses outweighed the gains in both years. In Nyangera, considerably higher seepage loss was observed in Season 3 than in Season 2. The total net seepage loss in Nyangera was four times higher, whilst the length of the functional period decreased by nearly half compared to Kusa. Precipitation and evaporation were the main driving forces for the pond water balance after flood recession, contributing over 93% of water gains and about 71 % of losses (Figure 3.8). The gains through runoff were negligible. There was considerably higher seepage loss in 2004, probably due to drought resulting in a rapid decline in the groundwater level in the wetland. Another additional loss of water from the ponds occurred through abstraction for vegetable irrigation on the raised bed gardens and other domestic uses, contributing up to 9% loss in pond water in Nyangera in Season 2.

Table 3.5: Seasonal water budget for Fingerponds (cm)

Variable	2003-2004		2004-2005	
	Nyangera	Kusa	Nyangera	Kusa
Duration of season (days)	243	184	123	184
Gains				
Precipitation	65.44	32.06	24.9	46.56
Runoff	5.21	2.67	2.49	3.10
Total	**70.65**	**34.73**	**27.39**	**49.66**
Losses				
Evaporation	76.20	73.72	52.44	73.37
Seepage	14.20	6.41	60.61	30.59
Abstraction	9.31			
Total	**90.93**	**94.86**	**113.05**	**103.96**

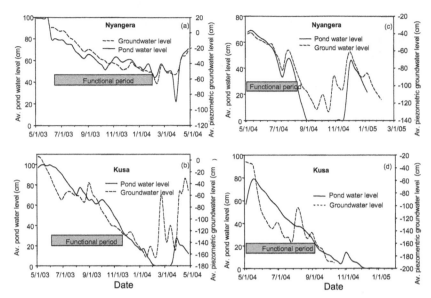

Figure 3.7: Observed groundwater and pond water level variations during two Fingerponds seasons in 2003-2004 and 2004-2005

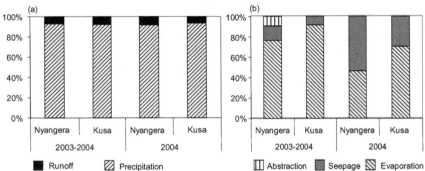

Figure 3.8: Water budget components of Finger ponds as a fraction of overall water gains (a) and overall losses (b)

Simulation of pond water level variation

The predicted pond water levels closely correlated with observed values (Figure 3.9 a and b). The predicted duration of the functional period for the two scenarios (1-variation in the maximum pond volume; and 2 - varying weather conditions) are shown in Figure 3.10. Under normal weather conditions and with an annual precipitation of about 1000 mm (typical of the region around Lake Victoria, Kenya), the pond functional period would last about 6 months (Figure 3.10 a, Scenario A 1). An increase in average depth by 0.5 m compared with the average depth of about 1 m in the experimental pond increases the potential functional period of the pond by about 2½ months (Scenario A 2). On the other hand a low water volume reduces the functional period by about 1½ months (Scenario A 3). Dry weather conditions with low precipitation shorten the functional period by nearly one month (Figure 3.10b,

Scenario B 2) whilst accelerated seepage in a situation where there is rapid decline in groundwater may further shorten the culture period by an additional two months (Scenario B 3).

The initial floodwater volume harvested, and the weather conditions after flood recession play an important role with regard to pond water supply and consequently determine the duration of the functional period.

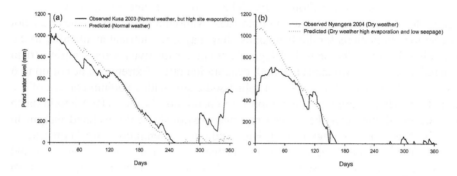

Figure 3.9: Model predicted and observed pond water level variation during the Fingerponds season

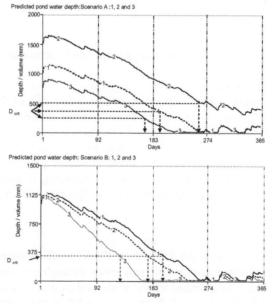

Figure 3.10: Simulated pond water level for (a) varying maximum pond volumes and (b) different hydrological conditions

Discussion

Soil properties and their influence on Fingerponds performance

Clay soils predominated at both sites indicating suitability for pond construction and aquaculture (Boyd et al., 1999). The soil organic matter (6-8%) is considered to be moderately high (Aguillar-Manjarez and Nath, 1998). This is common in wetlands due to accumulation of plant biomass (Ashley et al., 2004) and may lead to a reducing environment at the pond sediments. On the other hand, oxidation of such sediments may act as a carbon source. The cation exchange capacity (CEC) was high with over 40 meq/100 g soil in both sites. Although high CEC values may not pose serious limitations for the fish ponds, they may limit nutrient availability to the horticultural crops. The soils had high electrical conductivity values attributed to mineralization in the wetlands and high evaporation rates. Extreme values of up to 6 mS cm^{-1} in Kusa are attributable to a saline-sodic soil with a sodium content of 3.2 meq/100 g. High sodium concentrations are uncommon in the Lake Victoria basin but patches of such soils have been reported previously at the wetland margin in Kusa by Mati and Mutunga (2003). High sodium concentrations are known to be toxic to crops and may lead to poor performance (Abrol et al., 1988; Rengasamy and Olson, 1991). In Kusa the sodic-saline soils limited the overall Fingerpond system productivity. Nitrogen and phosphorus concentrations of 75 and 25 ppm respectively indicated low values in the bottom sediments of our new ponds and could explain partly the observed initial low fish yields obtained in un-manured Fingerponds (Kipkemboi et al., 2006). Over 250 ppm available nitrogen and 60 ppm available phosphorus in pond soils are needed for good fish production (Benerjia, 1967; Boyd and Bowman, 1997).

The role of natural events in water supply and the functioning of Fingerponds

Flooding and water supply
Field observations during this study indicated that inundation of the littoral wetlands is uncertain. In 2005, the water level in Lake Victoria declined to its lowest below the normal 0.5 metre oscillation (Figure 3.11). As a result, even the heavy rains in the catchment basin could not raise levels to the normal seasonal peak and consequently, the Fingerponds did not flood at the beginning of the wet season (April/May) in 2004. Apart from water supply, flooding is also essential for fish migration through the swamp and the stocking of the ponds (Petr, 2000; Kipkemboi et al., 2006).

Pond water balance
After flood recession, the water balance in Fingerponds is determined by natural gains through precipitation and runoff and losses through evaporation and seepage. The daily water balances obtained in this study and in other aquaculture studies are shown in Table 3.6. The results indicate that rainfall, a main source of water after flooding, is low at less than 3 mm day^{-1}. Most areas around Lake Victoria have been classified as agro-ecological zone IV, which is in the sub-humid to semi arid zone receiving an annual rainfall of between 600-1100 mm y^{-1} (FAO,1996).

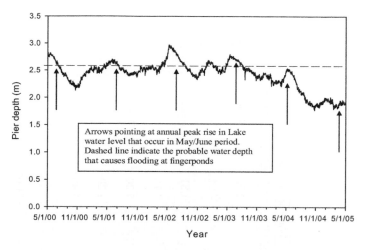

Figure 3.11: Lake Victoria water level 1999-2005 from measurement at pier located at S 00°
08′ E 34° 74′. (Source: LVEMP/GOK, 2005; Lake Victoria Environment Management
Project/Ministry of Water and Irrigation)

Highest rainfall usually occurs from April to May, but the functional period usually
starts in May/June after flooding of the wetlands. Thus, part of the potential water
gain through precipitation is outside the Fingerpond functional period. The
estimated gain due to runoff was insignificant and ranged between 0.09 and 0.21
mm day^{-1}, lower than most observations from other studies. Evaporation contributed
about 22 mm loss per week. Although this is low compared to the 42 mm weekly
loss reported in fish ponds in Honduras (Green and Boyd, 1995), it still had a
negative effect on the functional period of the Fingerponds. Seepage of up to 7.58
mm day^{-1} in Nyangera during the dry season of 2004 is considered only moderate
(Yoo and Boyd, 1994; Boyd and Gross, 2000). However in Fingerponds this can
contribute to a significant overall water loss considering the fact that the ponds do
not have regulated inflows. The implication of this for pond management is that
there is a high probability of ponds drying up prematurely should the negative fluxes
be accelerated by climatic condition. Seepage and evaporation rates were
comparable to other pond aquaculture studies, however, there was a high variability:
Nyangera showed consistently higher seepage rates whilst comparatively higher
rates of evaporation occurred in Kusa.

In pond aquaculture, water supply and particularly the initial filling is usually a
daunting task since it constitutes the largest single input requirement during the
season. For conventional ponds, the water supply may come from regulated surface
water channels or wells. Estimation of water requirement is critical in understanding
water supply and pond management. Although, the water supply in Fingerponds
depends on natural events such as floods, rainfall and groundwater exchange, an
hypothetical estimate of water requirement assuming a regulated flow can be used to
show the importance of the various sources of water supply. An estimate of the
water requirement for Fingerponds was made based on the hydrological equation of
Boyd and Gross (2000). The proposed guideline for Fingerponds is that individual
households may construct ponds of about 200 m^2 similar to our experimental
Fingerponds.

Table 3.6: Mean daily water balances from selected pond aquaculture studies, compared with Fingerponds

Source	Pond type and location	Method	Water gains (mm/day)			Water losses (mm/day)		
			Rainfall	Runoff	Inflow	Evaporation	Seepage	Outflow
This study	Fingerponds, Kenya	Observed	2.69	0.21	None	3.14	2.26	None
This study	Fingerponds, Kenya	Observed	1.13	0.09	None	2.60	4.91	None
This study	Fingerponds, Kenya	Observed	2.02	0.18	None	4.26	1.81	None
This study	Fingerponds, Kenya	Observed	2.53	0.17	None	3.99	2.16	None
Braaten and Flaherty (2000)	Inland shrimp, Thailand	Observed	5.20	0.30	9.40	3.10	5.20	7.00
Briggs and Funge Smith (1994)	Coastal shrimp, Thailand	Estimated	5.00	0.40	42.50	6.00	1.20	40.70
Nath and Bolte (1998)	Inland fish, Thailand	Modelled	6.10	0.30	2.40	4.80	3.90	0
Nath and Bolte (1998)	Inland fish, Honduras	Modelled	3.40	0.40	5.40	3.20	4.80	0
Teichert-coddington et al. (1988)	Inland Tilapia, Alabama	Observed	9.00	13.00	13.00	4.00	31.00	0
Boyd (1982)	Inland catfish, Honduras	Observed	2.80	0.10	8.90	3.60	7.80	0
Green and Boyd (1995)	Inland fish, Honduras	Observed	2.70	0.10	5.40	6.00	2.70	0

Modified from Braaten and Flaherty (2000)

The average depth at the middle of the pond should be about 1.5 m and assuming the ponds will dry up once a year, the average change in water storage is 150 cm. When the pond does not flood the water input will be governed by direct precipitation, groundwater seepage inwards and runoff from the raised-bed gardens. The functional period for Fingerponds should be at least six months (or 184 days) to allow juvenile migrant fish to grow to table size. Based on these assumptions, the estimated water requirement was derived as:

$$Q = (E+S)-P \pm \Delta H \tag{3.7}$$

where Q is the net water requirement, E is evaporation, S is seepage loss, P is precipitation and ΔH is change in storage. Using average rates of 0.35, 0.28 and 0.21 cm day^{-1}, a total evaporation, seepage and precipitation of 64.4, 51.5 and 38.6 cm, respectively was obtained. Substituting the above variables into equation 3.7 yields a water demand of 227.3 cm. The initial filling by flood provides a supply of 150 cm while the remaining 77.28 cm relies on natural gains mainly through precipitation and inward seepage (S_i) during the season. This indicates that flooding is important for the functioning of Fingerponds.

The effect of water supply on Fingerponds functional period
Water availability plays an important role in the determination of the Fingerponds functional period. During the dry season the water level may decline to a critical depth, beyond which the fish will be subjected to increased predation by piscivorous birds, mammals (otters), reptiles (Monitor lizards) and stress due to deteriorating water quality. In Season 2, the length of the functional period was 7.6 and 6.3 months in Nyangera and Kusa, respectively. It was shorter in Season 3 with only 3 and 4 months in Nyangera and Kusa, respectively. Dry weather was responsible for the accelerated water loss from the ponds and hence shortening of the functional period. If manuring is not adjusted accordingly, a rapid deterioration of water quality will ensue. In the dry season, the ponds may desiccate completely. Drying of the ponds allows aerobic decomposition of the bottom organic material which enhances the availability of nutrients for algal production. Drying also allows desludging, which increases the lifespan of the ponds and makes sediment nutrients available for crop production. On the other hand, drying has a negative impact on pond nutrient dynamics as nitrogen may be lost through volatilisation of ammonia (Boyd et al., 2002).

A disadvantage of the reliance on natural events for water supply and fish stocking is the uncertainty and lack of control by the farmer. The ponds may not flood at all and stay dry at the beginning of the season or there may be unexpected flooding which leads to a short Fingerponds season (as observed in December 2002 when an unexpected flood in both sites led to a season of only three months). This introduces an element of risk associated with Fingerponds.

Conclusion

Wetland physical and chemical soil characteristics vary between wetlands and although generally suitable for aquaculture, some sites may present challenges for the performance of Fingerponds. New pond sediments may be deficient in essential nutrients and require initial manure supplements. Flooding is seen as a limitation to aquaculture (Aguilar-Manjarez and Nath, 1998); nevertheless this study demonstrates that the moderate annual rise in water level can be used to enhance the fisheries in wetlands.

Fingerponds provide an opportunity for harvesting floodwater, which can then be used for aquaculture and irrigation hence enhancing wetland food supply. Flooding of the wetland is not only critical for water supply but also for saturating the soil at the beginning of the Fingerpond season and thereby moderating losses from seepage. After flood recession, precipitation is the principle source of water gain, while evaporation is mainly responsible for the loss.As flooding is a natural event, the dependence on it for the initial filling of the ponds and natural fish stocking creates uncertainty. This is a major challenge for the functioning of Fingerponds systems.

The duration of the Fingerponds functional period after flood recession is determined by the initial floodwater volume harvested and the prevailing weather conditions. Deepening the ponds to at least to an average depth of 1.5 m enhances storage capacity and consequently lengthens the functional period of Fingerponds.

Acknowledgement

I wish to acknowledge the financial support from the European Union Fingerponds project Contract no. ICA4-CT-2001-10037. Additional funding for field work and equipment was provided by the International Foundation for Science, Stockholm, Sweden and Swedish International Development Cooperation Agency Department for Natural Resources and the Environment (Sida NATUR), STOCKHOLM, Sweden, through grant no W/3427-1. I wish to express my thanks to Jan C. Nonner of UNESCO-IHE for guidance on hydrology issues. Last but not least, I appreciate the cooperation from the Fingerponds Project research team and the local communities at the experimental sites, which contributed to the success of this study.

References

Abrol, I.P., Yadov, J.S.P., Massoud, F.I., 1988. Salt affected soils and their management. FAO Soils Bulletin no. 39, Food and Agricultural Organization of the United Nations, Rome.

Aguilar-Manjarez, J., Nath, S.S., 1998. A strategic reassessment of fish potential in Africa. *CIFA Technical Paper No. 32*, FAO, Rome, 170 pp.

Ashley, G.M., Mworia, J.M., Muasya, A.M., Owen, R.B., Driese, S.G., Hover, V.C., Renaut, R.W., Goma, M.F., Mathai, S., Blatt, S.H., 2004. Sedimentation and recent history of a freshwater wetland in semi-arid environment: Loboi Swamp, Kenya, East Africa. Sedimentology 51, 1301-1321.

Benerjia, S.M., 1967. Water quality and soil conditions of fish ponds in some states of India in relation to fish production. Indian Journal of Fisheries 14, 115-144.

Boyd, C.E., Bowman J. R., 1997. Pond bottom soils. In Egna, H.S., Boyd, C.E. (ed.), Dynamics of Pond Aquaculture, CRC Press, Boca Raton, Florida, pp. 135-162.

Boyd, C.E., Queiroz, J.,Wood, C.W., 1999. Pond soil characteristics and dynamics of soil organic matter and nutrients. In: K. McElwee, D. Burke, M. Niles and H. Egna (Eds.), Sixteenth Annual technical Report. Pond Dynamics /Aquaculture CRSP, Oregon State University, Corvalis, Oregon, pp. 1-7.

Boyd, C.E., 1982. Hydrology of small experimental fish ponds at Auburn, Alabama. Transactions of the American Fisheries Society 111, 638-644.

Boyd, C.E., Gross, A. 2000. Water use and conservation for inland aquaculture ponds. Fisheries Management and Ecology 7, 55-63.

Boyd, C.E., Wood, C.W., Thunjai, T., 2002. Aquaculture pond bottom soil quality management. Aquaculture Collaborative Research Support Program, Oregon State University, Corvallis, Oregon, 41 pp.

Braaten, R.O., Flaherty, M., 2000. Hydrology of inland brackish water shrimp ponds in Chachoengsao, Thailand. Aquacultural Engineering 23, 295-313.

Briggs, M.R.P., Funge-Smith, S.J., 1994. A nutrient budget of some intensive marine shrimp ponds in Thailand. Aquaculture and Fisheries Management 25, 789-811.

Denny, P., Kipkemboi, J., Kaggwa, R. and Lamtane, H., 2006 The potential of Fingerpond systems to increase food production from wetlands in Africa. International Journal of Ecology and Environmental Sciences 32(1), 41-47.

Denny, P., 1989. Wetlands. In: Strategic resources planning in Uganda. UNEP Report IX 103 pp.

Denny, P., Turyatunga, F., 1992. Ugandan Wetlands and their Management. In: E. Maltby, P.J. Dugan and J.C. Lefeuver (Eds.). Conservation and Development: The sustainable Use of Wetland Resources. Proceedings of the Third International Conference, Rennes, France, 19-23, September 1988. IUCN, Gland, Switzerland. xii, pp. 77-84.

FAO, 1996. Agro-ecological zoning: Guidelines. Soil Bulletins no. 73, Food and Agriculture Organisation of the United Nations, Rome.

FAO, 1998. Wetland characterization and classification for sustainable agricultural development. Food and Agriculture Organisation of the United Nations, Sub-Regional office for East and Southern Africa, Harare.

Fernando, C.H., Halwart, M., 2000. Possibilities for the integration of fish farming into irrigation systems. Fisheries Management and Ecology 7, 45-54.

Gee, G.W., Bauder, J.W., 1986. Particle size analysis. In: Klute, A. (Ed.) Methods of soil analysis; Part 1. Physical and mineralogical methods, 2nd edition, American Society of Agronomy, Madison, Wisconsin USA. pp. 383-408.

Green, B.W., Boyd C.E., 1995. Water budgets for fish ponds in the dry tropics. Aquacultural Engineering 14(4), 347-356.

Halwart, M., van Dam, A.A., 2006. Integrated irrigation and aquaculture in West Africa: concepts, practices and potential. Food and Agriculture Organization of the United Nations (FAO), Rome, 181 pp.

Kapetsky, J.M., 1994. A strategic assessment of warm-water fish farming potential in Africa. *CIFA Technical Paper,* No. 27. Rome, FAO, 67 pp.

Kelly, A. M., Kohler, C.C., 1997. Climate, site and pond design. In: Egna, H.S., Boyd, C.E. (Eds.), Dynamics of pond Aquaculture, CRC press, Boca Raton, Florida pp 109-133.

Kipkemboi, J., van Dam, A. A., Denny, P., 2006 Biophysical suitability of smallholder integrated aquaculture-agriculture systems (Fingerponds) in East Africa's Lake Victoria freshwater wetlands. International Journal of Ecology and Environmental Sciences 32(1), 75-83.

LVEMP/GOK, 2005. Lake Victoria Environment Management Project (LVEMP) database/Government of Kenya (GOK), Ministry of Water and Irrigation unpublished data.

Mati, B.M., Mutunga, K., 2003. Integrated soil fertility and management assessment of the Kusa profile, Lake Victoria basin. Technical Report for Regional Land management Unit (Relma), Kusa pilot project, 89 pp.

McCartney, M.P., Musiyandima, M., Houghton-Carr, H.A., 2005. Working wetlands: classifying wetland potential for agriculture. Research Report 90, International Water Management Institute (IWMI), Colombo, Sri Lanka.

Nath, S.S., Bolte, J.P., 1998. A water budget model for pond aquaculture. Aquacultural Engineering 18, 175-188.

Okalebo, J.R., Gathua, K.W., Woomer, P.L., 2002. Laboratory methods of soil and plant analysis; A working manual. 2nd edition, SACRED Africa, Nairobi, 128 pp.

Petr T., 2000. Interactions between fish and aquatic macrophytes in inland waters; a review, FAO Fisheries Technical Paper. No. 396, Rome, 186 pp.

Rengasamy, P., Olson, K.A., 1991. Sodicity and soil structure. Australian Journal of Soil Resources 29, 935-952.

Spieles, D.J., Mitsch, W.J., 2000. The effects of season and hydrologic and chemical loading on nitrate retention in constructed wetlands: a comparison of low and high nutrient riverine systems. Ecological Engineering 14, 77-91.

Teichert-Coddington, D.R., Stone, R.P., Phelps, R.P., 1988. Hydrology of fish culture ponds in Gualaca, Panama. Aquacultural Engineering 7, 309-320.

US Soil Conservation Service, 1972. Hydrology. In: SCS National Engineering Handbook, Section 4, US Soil Conservation Service, Washington DC, 400 pp.

Welcomme, R.L., 1975. The fisheries ecology of African floodplains. CIFA Technical Paper. No. 3, FAO, Rome 54 pp.

Welcomme, R.L., Bartley, D.M., 1998. An evaluation of present techniques for the enhancement of fisheries. In: T. Petr (ed.) Inland fishery enhancement. Paper presented at the FAO/DFID expert consultation on inland fishery enhancements, Dhaka, Bangladesh, 7-11 April, 1997.

Yoo, K.H., Boyd, C.E., 1994. Hydrology and water supply for pond aquaculture, New York, Chapman and Hall, 483 pp.

Zhang, L., Mitsch, W.J., 2005. Modelling hydrological processes in created freshwater wetlands: an integrated system approach. Environmental Modelling & Software, 20, 935-946.

Chapter 4

The effects of livestock manure on nutrient dynamics, water quality and fish yields in seasonal wetland fish ponds (Fingerponds) at the edge of Lake Victoria, Kenya

Abstract

This chapter reports the results of pond aquaculture dynamics of experimental integrated aquaculture ponds (Fingerponds) at the Lake Victoria wetlands in Kenya. Fingerponds are self-stocked flood-based fish ponds at the wetland margin. The overall aim of the study was to develop and evaluate low-cost production systems to enhance the wetland fishery at subsistence level. The ponds were stocked adequately with diverse fish species at densities of 3 to 10 fish m^{-2}. The effects of livestock manure applications on nutrient dynamics, water quality and fish yields were studied. There was no observable adverse effect of manuring on pond water quality. Regression analysis indicated that site, pond management (manuring) and the environmental and climatic variables explained a large part of the variation in NH_4-N, NO_3-N and total nitrogen concentrations with adjusted r^2 of 0.64, 0.70 and 0.65, respectively. The explained variance for o-PO_4 and total phosphorus was 58% and 61%, respectively. Manuring increased the total phosphorus concentration in the sediment but only had marginal effects on total nitrogen. The chlorophyll a concentration was higher in manured ponds, reaching an average of 150 μg l^{-1} compared to an average of 27 μg l^{-1} in un-manured ponds. The net fish yields were highly variable between sites and seasons and ranged from 402 to 1069 kg ha^{-1}, the data showing that manuring was advantageous. The duration of the culture period, site variability and manuring explained 82% of the variation in fish yields. Careful fertilization of the ponds with livestock manure can be used to improve fish yields in Fingerpond systems.

Key words: Lake Victoria wetlands, Kenya, aquaculture, natural fish stocking, fish yields, Fingerponds

Publication based on this chapter:

Kipkemboi, J., A.A. van Dam, C.M. Kilonzi, N. Kitaka, J.M. Mathooko, P. Denny. The effects of livestock manure on nutrient dynamics, water quality and fish yields in seasonal wetland fish ponds (Fingerponds) at the edge of Lake Victoria, Kenya. Aquaculture (submitted).

Introduction

Fish is nutritionally important and provides 20% or more of animal protein to the majority of the population in sub-Saharan African countries (FAO, 2004a). For many rural households living around the continent's water bodies, capture fisheries has been an important livelihood asset. However, environmental degradation and unsustainable exploitation over the recent years have threatened the sustainability of this natural resource. In Lake Victoria, East Africa, environmental degradation and over-fishing have led to the decline of the fishery (Odada et al., 2004) and generally, in sub-Saharan Africa, the per capita fish supply has fallen (FAO, 2005): this could be restored through development of sustainable aquaculture production.

The biophysical potential for fish farming in Africa is demonstrated by Aguilar-Manjarez and Nath (1998) and Li and Yang (1995) have shown that littoral wetlands could be used for this purpose. Traditionally, African floodplains play an important role in seasonal fishery for rural communities and this can be enhanced (COFAD, 2002) but their potential for aquaculture is yet to be realized. The growth of conventional aquaculture systems developed earlier in the region were hampered by inadequate inputs (especially supplemental feeds and fingerlings) as well as inappropriate technologies and weak research and extension (Machena and Moehl, 2001; FAO, 2004b). To overcome these challenges, there is a need to develop low-input production systems. Such systems should aim at integration into other farming activities to enhance the sustainability potential through synergy (Brummett, 1999; Jamu and Ayinla, 2003; Ofori et al., 2005, Halwart and van Dam, 2006). Smallholder aquaculture systems similar to the Asian integrated systems have been tried with considerable success in Malawi and parts of West Africa (Brummett and Noble, 1995; Prein et al., 1995).

This study investigated experimental integrated aquaculture–agriculture systems called "Fingerponds" in East Africa's Lake Victoria wetlands. Fingerponds are earthen ponds excavated in the fringe wetlands during the dry season. The excavated soil is spread around the ponds to create raised-bed gardens for vegetable production. The ponds resemble the natural flood pools traditionally used for wetland fish capture by local communities while the gardens are a continuation of the existing seasonal swamp margin vegetable patches. They are called "Fingerponds" because from a bird's eye view several of these narrow channel-like ponds look like "fingers" extending into the emergent macrophyte zone. Pond-stocking occurs naturally when wetland wild fish are carried by the floods and become trapped in the ponds as the floodwater recedes. The water supply in Fingerponds is un-regulated and is determined by natural gains and losses (Chapter 3 of this thesis). Fish culture and garden management start after flood recession. Manure from livestock and vegetable wastes are applied to the ponds to stimulate the production of natural fish food.

In earthen fish ponds, sediments act as nutrient sources (through sediment–water interface release) as well as sinks (through sedimentation) (Boyd and Bowman, 1997; Delincé, 2000). Therefore, they play an important role in nutrient cycling and pond productivity. As new pond soils are usually deficient in the limiting nutrients (nitrogen and phosphorus), fertilization stimulates primary production and can be used to manipulate the natural fish food availability and overall pond productivity (Yussoff and McNabb, 1989; Diana et al., 1991; Veverica et al., 2001). Animal manure is used as a nutrient source for aquaculture ponds (Wahby, 1974; Wohlfarth and Schroeder, 1979; Knud-Hansen, 1998) and in Africa, is environmentally attractive since it promotes synergy and efficient use of farm nutrient resources. Fertilization of ponds with organic manure has positive and negative effects on water quality and consequently on fish survival and growth (Lin et al., 1997). Prudent manure use is crucially important for Fingerpond systems as they lack regulated water flows. The objectives of this study were: (1) to assess the potential for natural fish stocking and fish density dynamics in experimental Fingerponds; and (2) to investigate the effects of livestock manure on pond water quality parameters, sediments nutrients, and on fish growth and yields.

Materials and Methods

The study was carried out at the Fingerponds experimental sites in Nyangera and Kusa at the shores of Lake Victoria, Kenya (Chapter 1). Site selection for the experimental ponds was based on the regularity of flooding during the annual water rise in the emergent macrophyte zone of the wetland. This ensured that flooding and the subsequent migration of wild fish into the ponds occurs during the natural inundation of the wetland. Four rectangular earthen ponds measuring 24 m long by 8 m wide and a sloping bottom of 2 m at the deep end and 1 m at the shallow end were constructed in each of the study sites between April and October, 2002. The digging was manual by the local communities at the respective sites. The construction was carried out during the dry season when groundwater was supposedly lowest in the wetland to avoid difficulties of groundwater seepage into the ponds while digging. The ponds were filled with floodwater from the lake shortly after the onset of the long rains.

Flooding, Fingerponds functional and fish culture periods

Flooding marks the beginning of the events in Fingerponds. After the completion of construction in October 2002, the ponds were flooded by unexpected floods in December followed by a short functional period from January to March 2003. The functional period is defined as the period between the isolation of the ponds and the harvest of the fish when the water level has dropped to the critical depth (defined in Chapter 3). In 2003, flooding of the ponds occurred in May and lasted for two weeks in Kusa, but was extended until July in Nyangera. The 2003/2004 season was regarded as a normal Fingerponds season and lasted 227 days in Nyangera and 190 days in Kusa. The fish culture period after re-distribution of the fish was 154 days in Nyangera and 146 days in Kusa. Usually the ponds would be expected to dry up between December and February (the driest period of the year). However, in Nyangera the ponds did not dry completely in the 2003/2004 dry season while in Kusa they dried completely in January. When the long rains resumed in April-May 2004, the lake flooding was inadequate but the ponds filled with direct precipitation and infiltration from groundwater. Since the Nyangera sites did not dry up, there was a carry-over of fish in the ponds from the previous season, whilst in Kusa they did not become stocked from flooding. The Kusa ponds were then stocked by the carry-over of stock from the Nyangera site. A detailed description of the pond hydrology during the experimental period is presented in Chapter 3.

Fish stock assessment and identification

After flood recession, when the ponds became isolated, the fish natural stocking density was determined by seining through the ponds with a 6.5 mm mesh size seine net. Fish stocks were identified with the help of taxonomic keys (Van Oijen, 1995). Fish sizes greater than about 5 cm in total length (TL) were measured, counted and weighed to the nearest 1 g while the smaller fish were batch weighed and counted. Since complete removal through seining is not possible, the total population was estimated through extrapolation of depletion in numbers in repeated seines using Microfish 3.0 software (Van Deventer and Platts, 1985). The density per pond was determined and in the case of un-even stocking, the fish were re-distributed to achieve nearly balanced stocking densities.

Pond manuring and water quality monitoring

During the fish culture period, the ponds were fertilized with livestock *boma* (animal enclosure) manure obtained from the adjacent villages. The manure was mainly from cattle but was often mixed with sheep and goat faeces as it is common practice to put the animals in the same enclosures at night. The total nitrogen and total phosphorus concentration in manure samples averaged 19.67 ± 1.45 mg g^{-1} and 3.28 ± 0.19 mg g^{-1} in Nyangera and 18.35 ± 0.78 mg g^{-1} and 3.25 ± 0.04 mg g^{-1} in Kusa, respectively. Two application rates of 1250 (low) and 2500 (medium) kg ha^{-1} per 2 week intervals of dry manure were used. In each site, two ponds received the medium manure application, one pond was used for low manure and one pond remained un-manured as a control. In Nyangera, manure was broadcasted daily on the shallow section of the pond whilst in Kusa manure was soaked in a gunny bag and placed in one corner of the pond at the shallow end. After a fortnight, the composted manure in the gunny bag was emptied and spread on the shallow end section before filling the bag with fresh manure.

Pond water quality variables were monitored monthly during the fish culture period. Physical parameters were measured on site using a Jenway model 350 pH meter (Essex, U.K) for pH, a Jenway model 470 conductivity/TDS meter (Essex, U.K) for electrical conductivity, and a portable WTW (Wissenchaftlich-Technische Werkstätten) dissolved oxygen meter model 330i (GmbH & Co. KG, Weiheim, Germany) for dissolved oxygen. Pond turbidity was measured daily using a secchi disk (Lind, 1979).

Water samples for soluble nutrients were collected and filtered into acid-washed polythene bottles immediately after sampling. Filtering was carried out on site using a disposable syringe to push the water sample through GF/C 47 mm glass fibre filters (Whatman, U.K.) secured by Swinnex filter holders. Unfiltered samples were collected for total nitrogen, total phosphorus, total suspended solids, biochemical oxygen demand (BOD$_5$) and chlorophyll *a* determinations. All samples were stored immediately in cool boxes and transported to the laboratory under ice. In the laboratory, BOD$_5$ incubation was carried out immediately and the filtered samples were analyzed for soluble nutrients. The analysis for orthophosphate (o-PO$_4$), total phosphorus (TP), ionized ammonia (NH$_4$-N), nitrate-nitrogen (NO$_3$-N), total nitrogen (TN), total suspended solids (TSS), alkalinity as CaCO$_3$ and biochemical oxygen demand (BOD$_5$) followed standard methods (APHA, 1992; 1995). Determination of chlorophyll *a* concentration followed the methodology proposed by Pechar (1987). Incompletely analysed samples were stored in a refrigerator at 4°C for further processing on the following day.

Sediment analysis

Three replicate pond sediment samples were collected from the deep, middle and shallow end of the ponds (coded as zones A, B and C respectively) using a core sampler. The cores were cut into three slices of 3 cm thickness from the surface, i.e. 0-3 cm, 3-6 cm and 6-9 cm deep. Air-dried samples were analyzed for total phosphorus and total nitrogen following methods outlined by Okalebo et al. (2002). Sample digestion was based on wet acid oxidation using Kjeldhal apparatus. Samples for total nitrogen were distilled for ammonia using a Kjeltec model 2200 Auto distillation apparatus and titrated with hydrochloric acid. Total phosphorus samples were determined calorimetrically using the ascorbic acid method.

Fish harvesting

Fish harvesting was carried out at the end of the season, before the ponds dried out completely. Repeated seines (in most cases a minimum of 8 times) using a 6.5 mm mesh size net were carried out to ensure that most of the fish were caught. Fish sizes greater than 5 cm were weighed and measured for total length (TL) while small fish were batch weighed and counted. The net fish yield (NFY) was calculated as the difference between the total weight of fish at the beginning of the season and the total final weight at harvest.

Data analysis

Statistical analysis was performed using SPSS 11.0 for Windows (SPSS Inc., Chicago, USA). Prior to statistical analysis the data for parametric tests were checked for normality and when violation was observed, appropriate transformations were applied. A t-test was used to compare the nutrient concentration between manured and un-manured ponds.

Multiple linear regression was used to assess the effect of manuring on sediment nutrient concentration; and the effect of location (sites), pond management (total amount of manure added) and environmental/climatic variables (pH, water temperature, rainfall, dissolved oxygen concentration and water depth) on water quality (pond water nutrients: NH_4-N, NO_3-N, TN, o-PO_4, TP) and chlorophyll a. The partial regression coefficients (sign, value and the significance of the t-test) were used to infer the effects of various independent variables on the dependent variables whilst the F-test together with the adjusted r^2 were used to test the predictive power of the models. The Durbin-Watson coefficient was used to check for autocorrelation (van Dam, 1990).

Multiple regression analysis was also used to assess the spatio-temporal effects of manuring on sediment nutrients and organic matter. Ordinal and nominal data were transformed into sets of dichotomies (dummy variables). The dependent variables used were sediment total nitrogen and total phosphorus whilst the independent variables were the dummy variables for the sampling dates, sites, pond manuring and for zones.

A repeated measures ANOVA was used to evaluate the effects of sampling date, zone and depth on sediment nitrogen and phosphorus during the fish culture period in 2003. The time/sampling date was used as the within-subject factor while depth and zone were used as the between-subject factors. For this ANOVA, only the data from the Kusa site, where adequate replicate samples were available, were used.

The relationship between the net fish yield as dependent variable and manuring, culture period duration and sites was also evaluated using multiple linear regressions. A stepwise regression was applied to determine the variation accounted for by individual factors.

Results

Fish species composition and densities

The fish migrants into the Fingerponds through flooding consisted of several species, resulting in a polyculture. The dominant group consisted of three species of tilapia: *Oreochromis niloticus* (Linnaeus, 1758), *O. variabilis* (Boulenger, 1906) and *O. leucostictus* (Trewavas, 1933). The typical wetland fish species *Protopterus aethiopicus* (Heckel, 1906) and the small littoral zone fish *Ctenophoma muriei* (Boulenger, 1906) and *Aplocheilichthys* sp. were present but in small numbers. A variable occurrence of *Clarias gariepinus* and *Haplochromis* spp. were observed in both study sites (Figure 4.1).

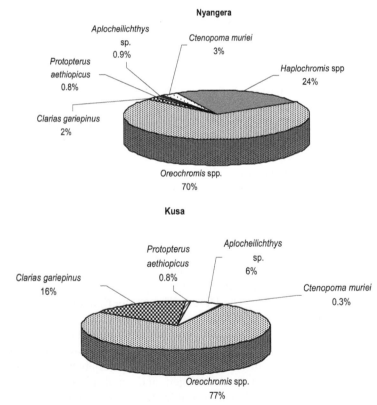

Figure 4.1: Fish species composition (% of the total number of fish in 4 ponds after flood recession) at two experimental flood-stocked Fingerponds sites (Nyangera and Kusa) near Lake Victoria in 2003.

The post-flood recession fish stock assessment revealed that the stocking densities varied between sites (Table 4.1). A fish census after the May 2003 floods revealed moderate stocking with an average of 3 fish/m^2 in Kusa while Nyangera had higher densities of 11 fish/m^2. To achieve higher growth rates for the tilapia their initial densities were reduced. In Nyangera, the densities were reduced to less than 1 fish per m^2 while in Kusa, where the initial stocking densities were moderate, fish were redistributed to obtain more balanced densities between ponds. After about 5 months, the average fish density in Nyangera had increased to an average of 23 fish/m^2, while in Kusa it declined to 1 fish/m^2.

Table 4.1: Fish stocking densities at the two experimental Fingerponds sites in Kenya in 2003. The initial densities after stocking by natural flood, manipulations through redistribution and final densities.

Site	Pond	Fish densities (individuals m^{-2})		
		Initial	On re-distribution	Final
Nyangera	1	9.77	0.69	25
	2	9.53	0.70	21
	3	12.58	0.70	19
	4	12.15	0.88	27
Kusa	1	3.41	3.1	0.51
	2	3.54	3.1	0.79
	3	3.70	3.1	1.74
	4	0.31	3.1	1.01

Water quality parameters

Table 4.2 summarizes the physical and chemical parameters in manured and un-manured ponds. The water temperature during the fish culture period ranged between 24 and 28°C in Nyangera and between 23 and 31°C in Kusa. In both sites, pH ranged from neutral to moderately alkaline (pH 9.00). The striking observation was the elevated electrical conductivity of greater than 4 mS cm^{-1} at the beginning of the culture period and the gradual increase as the pond water volume decreased during the dry season. Higher values of up to 10-12 mS cm^{-1} were measured in the dry season of 2004.

The manured ponds generally had higher nutrient concentrations than the un-manured ponds. Pooled data from manured and un-manured ponds showed that pond fertilization had a significant effect on electrical conductivity, NH_4-N, NO_3-N, total nitrogen, o-PO_4, total suspended solids (TSS), secchi disk depth (SDD), dissolved oxygen, BOD_5, and chlorophyll a (t-test, P<0.05). Total alkalinity ($CaCO_3$) did not differ significantly (P>0.1). Ammonium ion concentration was higher in manured ponds, reaching over 1 mg l^{-1} although no significant difference was observed between manured and un-manured ponds (P>0.05). The TSS in un-manured ponds averaged 190.0 ± 30.1 mg l^{-1} in Kusa, four times higher than 44.7 ± 3.0 mgl^{-1} observed in the same treatment in Nyangera. The TSS differed significantly between sites, even in un-manured ponds, implying that this was due to site differences rather than the result of pond fertilization. This was also indicated by the higher turbidity and the resultant low SDD measured in Kusa compared to Nyangera. Considerably higher BOD_5 was observed in manured ponds in Kusa compared to Nyangera. The chlorophyll a concentration increased steadily, reaching an average of 150 μg l^{-1} in ponds that received 2500 kg ha^{-1} per 2 weeks in both study sites (Figure 4.2). The dissolved oxygen levels were variable during the day, dropping to 3 to 5 mg L^{-1} in the morning and reaching 12 mg L^{-1} in the late afternoon (Figure 4.3).

The multiple regression models, using the pond water nutrients as dependent variables and location, total manure load, pond water volume (depth), rainfall, pH, water temperature and dissolved oxygen as independent variables were significant (P<0.001). The explained variance for ammonium-nitrogen, nitrate-nitrogen and total nitrogen concentration was 64.5%, 69.8% and 64.7%, respectively. The regression coefficient for manure load was significant and positive in all cases indicating the positive effect of manure on water nutrient concentrations. The partial regression coefficient for temperature in the ammonium-nitrogen model was significant and negative. The regression models for o-PO$_4$ and total phosphorus were significant (P<0.001) with explained variance of 58.5 % and 61 %, respectively. In both cases, the partial regression coefficients for manure load were significant and positive, again indicating a positive effect of manuring on phosphorus concentration in the pond water. The N:P ratios were lower in fertilized ponds compared with the un-manured ones except in the low-manured pond in Kusa.

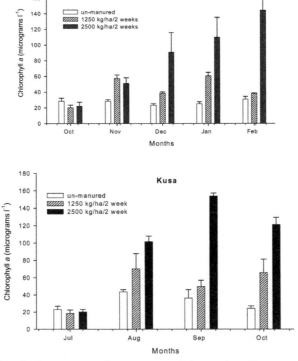

Figure 4.2: Chlorophyll *a* concentration un-manured, low and medium-manured Fingerponds in Nyangera and Kusa (mean ± standard error; figures are means of three replicates in one pond per site per sampling date [n=3], except for the medium manure treatment where three replicates from two ponds were taken [n=6]).

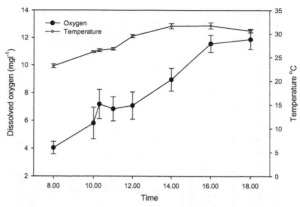

Figure 4.3: Dissolved oxygen and temperature variation during the day in manured
    Fingerponds in Kusa (values are means ± standard error of four measurements in manured
    ponds in 10th August 2004).

Pond sediment nutrients

Figure 4.4 shows the distribution of sediment nitrogen and phosphorus. The
distribution was nearly uniform throughout the entire pond bottom in Nyangera,
where manure was broadcasted from the shallow end. In Kusa, there was
accumulation of nutrients at the shallow end of the ponds where the manure was
applied and the leached and composted remains were spread. Table 4.3 summarizes
the regression results for total nitrogen and total phosphorus in pond sediments. The
regression analysis showed significant models with 42 % of the variance explained
for phosphorus (F= 9.49, P<0.001). Although the regression model for total nitrogen
was significant, the explained variation was only 12% (F = 2.45, P=0.03). The
regression coefficient for the dummy variable for sites for phosphorus was
significant and positive, indicating higher concentrations in Nyangera than in Kusa.
For nitrogen, the regression coefficient for site was not significant (P>0.1). In the 3-
6 cm sediment core depth, the regression models for total phosphorus and total
nitrogen were also significant with 44% and 10% of the variance explained,
respectively. Site differences were significant in both cases, but the coefficient for
total nitrogen was negative indicating lower concentration in Nyangera compared to
Kusa. The site coefficient for total phosphorus was positive, indicating a higher
concentration in Nyangera than in Kusa. In the 6-9 cm depth, the regression model
for total phosphorus was significant (F = 7.19, P < 0.001) while that of total nitrogen
was not significant even at P = 0.1. The explained variance for total phosphorus
declined to 36% whilst that of total nitrogen was negligible. The coefficients of total
phosphorus in zone A and B (deep and middle ends respectively) were negative for
all core depths although not significant except in zone B in the 3-6 cm depth.

The repeated measures ANOVA results are presented in Table 4.4. In un-
manured ponds the main effect of zone was significant (F = 4.89, P<0.05) and the
interaction between time and zone for sediment total nitrogen. However, for total
phosphorus only the effect of sampling date was significant (P<0.001).

Table 4.2: Summary of water quality parameters in manured and un-manured ponds in two locations near Lake Victoria, Kenya (Nyangera: September 2003 to February 2004; and Kusa: July to November 2004). Figures are mean (± standard error) of monthly measurements of physical and chemical parameters during the fish culture period (values in parenthesis indicate maximum values).

	Nyangera			Kusa		
	Un-manured	1250 kg ha⁻¹ per 2 weeks	2500 kg ha⁻¹ per 2 weeks	Un-manured	1250 kg ha⁻¹ per 2 weeks	2500 kg ha⁻¹ per 2 weeks
Temperature (°C) range	24.7 - 28.8	24.5 - 28.4	24.2 - 28.6	22.6 - 30	23.9 - 30.1	23.7 - 31.1
pH range	7.01 - 8.11	7.75 - 8.95	7.35 - 8.02	7.47 - 9.06	8.54 - 9.04	8.65 - 9.11
NH_4-N (mg l⁻¹)	0.31 ± 0.05 (0.74)	0.57 ± 0.15 (0.92)	0.73 ± 0.11 (1.33)	0.40 ± 0.06 (0.75)	0.45 ± 0.12 (0.80)	0.74 ± 0.16 (1.41)
NO_3-N	0.22 ± 0.03	0.35 ± 0.02	0.42 ± 0.03	0.21 ± 0.03	0.29 ± 0.07	0.36 ± 0.07
Total nitrogen (mg l⁻¹)	1.20 ± 0.05	1.71 ± 0.11	2.07 ± 0.13	1.25 ± 0.02	1.87 ± 0.30	2.29 ± 0.21
o-PO_4 (mg l⁻¹)	0.03 ± 0.01	0.06 ± 0.02	0.15 ± 0.03	0.05 ± 0.01	0.13 ± 0.04	0.32 ± 0.07
Total phosphorus (mg l⁻¹)	0.11 ± 0.02	0.48 ± 0.19	0.42 ± 0.09	0.18 ± 0.01	0.28 ± 0.07	0.67 ± 0.10
DIN:DIP ratio	4.82	1.92	2.73	12.2	0.87	3.44
Electrical conductivity (mScm⁻¹)	5.37 ± 0.24	4.74 ± 0.07	6.88 ± 0.21	4.62 ± 0.53	7.58 ± 0.71	5.97 ± 0.52
Total suspended solids (TSS) (mg l⁻¹)	44.7 ± 3.8	66.9 ± 13.0	108.2 ± 20.4	190. ± 30.1	250.0 ± 58.6	262.3 ± 26.9
Secchi Disk Depth (cm)	41 ± 3	32 ± 2	26 ± 1	16 ± 1	15 ± 1	14 ± 1
Alkalinity $CaCO_3$ (mg l⁻¹)	122.1 ± 20.3	131.4 ± 25.7	147.6 ± 25.5	144.7 ± 14.7	199.8 ± 25.4	195.7 ± 23.1
BOD_5 (mg l⁻¹)	2.92 ± 0.26	5.79 ± 1.18	7.67 ± 1.13	5.71 ± 0.94	12.13 ± 0.52	13.99 ± 0.59

DIN is total inorganic nitrogen and DIP is total inorganic phosphorus

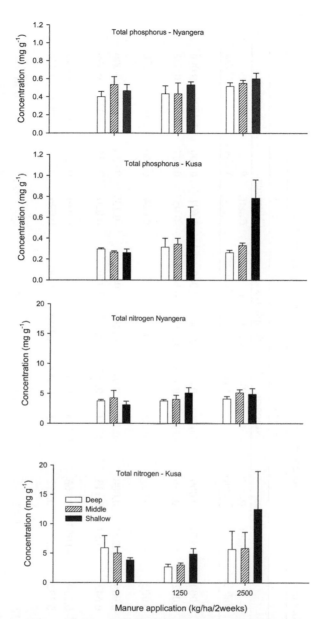

Figure 4.4: Average sediment total phosphorus and total nitrogen concentration in the different pond sections (deep, middle and shallow ends) at 0-9 cm core depth at two experimental Fingerponds sites in Kusa and Nyangera (mean ± standard error; n = 81 for un-manured and low manure; and n = 162 for medium manure).

Table 4.3: Multiple regression models for total nitrogen and total phosphorus in pond sediments in Nyangera and Kusa, Kenya.

Variables	0-3 cm				3-6 cm				6-9 cm			
	[a]Nitrogen		Phosphorus		[a]Nitrogen		Phosphorus		[a]Nitrogen		[b]Phosphorus	
	Coeff.	P-Values	Coeff.	P-Values	Coeff.	P-Values	Coeff.	P-Values	Coeff.	P-Values	Coeff.	P-Values
Regression coefficients												
Constant	2.155	0.000	1.476	0.000	2.078	0.000	0.286	0.000	1.691	0.000	1.392	0.000
Dummy for 1st sampling date	-0.168	0.002	-0.014	0.736	-0.148	0.194	0.015	0.547	-0.057	0.487	0.084	0.002
Dummy for 2nd sampling date	-0.254	0.002	-0.038	0.0315	-0.280	0.013	0.004	0.857	0.024	0.760	0.103	0.000
Dummy for site	-0.008	0.905	0.222	0.000	-0.190	0.044	0.149	0.000	-0.139	0.036	0.088	0.000
Dummy for manuring	-0.033	0.642	0.057	0.113	0.174	0.087	-0.019	0.390	0.020	0.744	-0.043	0.066
Dummy for zone A	-0.073	0.339	-0.053	0.166	0.136	0.207	-0.030	0.211	0.046	0.520	-0.13	0.619
Dummy for zone B	0.065	0.392	-0.039	0.312	0.145	0.183	-0.050	0.039	0.096	0.208	-0.009	0.776
ANOVA												
Adj R^2	0.12		0.42		0.10		0.44		0.021		0.36	
F-value	2.45		9.49		2.30		10.28		1.232		7.19	
Significance of F value	0.035		0.000		0.046		0.000		0.303		0.000	

Superscripts *a* and *b* imply square-root and logarithmic transformation of the data respectively

Table 4.4: Repeated measures ANOVA for total nitrogen and total phosphorus in pond sediments. Analysis based on pond sediment samples in Kusa, Kenya. Depth indicates sediment depth (0-3, 3-6, 6-9 cm), zone indicates location in the pond (shallow, middle and deep end) and date indicates weekly sampling date.

Interaction	d.f.	Un-manured				1250 kg/ha/2 weeks				2500 kg/ha/2 weeks			
		Nitrogen		Phosphorus		Nitrogen		Phosphorus		Nitrogen		Phosphorus	
		F	P	F	P	F	P	F	P	F	P	F	P
Between-subject effects													
Intercept	1	640.56	0.000	6499.86	0.000	637.0	0.000	554.65	0.000	798.47	0.000	786.62	0.000
Depth	2	1.03	0.379	1.07	0.367	3.06	0.074	0.30	0.744	0.32	0.732	0.06	0.942
Zone	2	4.89	0.022	0.99	0.394	5.73	0.013	6.03	0.011	0.14	0.866	7.24	0.005
Depth*zone	4	3.39	0.034	2.14	0.122	1.39	0.279	0.87	0.505	0.07	0.991	0.41	0.800
Within-subject effects													
Date	1	1.440	0.248	15.44	0.001	199.09	0.000	286.87	0.000	4.75	0.043	60.59	0.000
Date*depth	2	0.30	0.747	1.86	0.187	2.81	0.089	0.17	0.848	0.02	0.983	0.52	0.604
Date*zone	2	4.76	0.024	0.25	0.782	3.20	0.066	6.98	0.006	0.07	0.933	9.57	0.002
Date*zone *depth	4	0.96	0.456	1.19	0.351	1.12	0.378	0.77	0.560	0.27	0.892	1.80	0.176

In low-manured ponds, the main effect of zone and time were significant for both total nitrogen and total phosphorus while the interaction between sampling date and depth was significant for total phosphorus (P<0.05). In the medium-manured ponds, the effect of sampling date was significant for both parameters, and the zone and the interaction between the sampling date and zone were significant for total phosphorus (P<0.05). In general, the regression models indicated that manuring had more effect on sediment phosphorus concentration and only marginal for nitrogen. The repeated measures ANOVA revealed the effect of time on sediment nitrogen and phosphorus concentration indicating accumulation due to manuring. There were also significant differences between zones.

Fish growth and yields

Table 4.5 shows the variation in the fish densities during the season and fish productivity in un-manured and manured ponds. The experiments were conducted only in one season and the results are inconclusive. However, there was evidence of a high recruitment of juvenile fish through breeding within the ponds by tilapia in the Nyangera site and unexplained mortalities as observed in the Kusa site. The net productivity was considerably higher in Nyangera compared to Kusa.

Table 4.5: Comparison of fish densities and yields between un-manured and manured Fingerponds in Nyangera and Kusa, Kenya. Data based on culture periods between 29th September 2003 to 2nd March 2004 in Nyangera and from 22nd June to 15th November 2003 in Kusa.

Site	Treatment	Initial density (indiv. /m^2)	Final density (indiv. /m^2)	Net productivity (kg ha^{-1}day^{-1})	% Change in numbers during the 5-month culture period
Nyangera	Un-manured	0.70	19	3.14	2338
	Manured	0.76	24	6.01	3373
Kusa	Un-manured	3.1	0.51	0.11	-84
	Manured	3.1	1.18	1.96	-62

At the census after the first flood, the harvestable fish (> 15 cm or palm size) amounted to 38 kg ha^{-1} in Kusa and 115.4 kg ha^{-1} in Nyangera. During the re-distribution, the average fish yield cropped from 4 ponds in Nyangera was 199.3 ± 26.3 kg ha^{-1} (mean and standard deviation). In Kusa cropping was not carried out since the fish densities were moderate. The gross fish yields were highly variable between sites and seasons (Figure 4.5). In the first season when the ponds were not manured, the yields in Nyangera ranged from 161.6 to 520.8 kg ha^{-1} while in Kusa they varied from 132.2 to 258.8 kg ha^{-1}. The 2003-2004 period was regarded as a normal season with respect to the hydrological conditions (Chapter 3). During this period, the net fish yields obtained in manured ponds in Kusa averaged 401.9 ± 26.0 kg ha^{-1} (mean ± standard deviation). In Nyangera the net yields were higher and averaged 1068.6 ± 99.4 kg ha^{-1} (mean ± standard deviations) in manured ponds. Comparison of pooled net fish yield data between unmanured and manured ponds differed significantly (T-test = 2.20, P = 0.04), whilst the low and medium-manured ponds at the two sites showed no statistical difference (T–test =1.48, P>0.05, d.f. 8). A stepwise multiple regression analysis with the net fish yield as the dependent variable and site, fish culture period duration (months), and manuring revealed that the duration of the culture period accounted for 50% of the total variation while all the three factors explained 82% of the variation in fish yields (F = 23.81, P < 0.001, d.f. = 3,12).

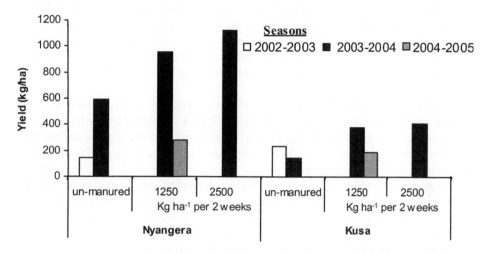

Figure 4.5: Fish yields versus manure application rates in the Nyangera and Kusa Fingerponds. Values are fish yields (kg ha[-1]) over three seasons with variable lengths; Nyangera (2.5, 7.6 and 2.6 months) and Kusa (2.8, 6.3, and 4.0 months) for 2002-2003, 2003-2004 and 2004-2005 seasons respectively.

Discussion

Fingerpond natural fish stocking
The natural fish stocking densities were adequate or in some cases excessive. In both experimental sites, the species composition was diverse, resulting in a polyculture system dominated by tilapia. This natural stocking has several advantages. One is that the different fish species can utilize the different niches available in the ponds resulting in a higher ecological efficiency. Secondly, predatory fish species such as *Clarias* sp. may be beneficial as they can help control the recruitment of tilapia. And thirdly, natural stocking implies zero costs for fingerlings which is a relief to the resource-poor households who are the target users of this technology.

The effects of pond manuring on pond water quality parameters
The water quality parameters were within the favourable range for fish culture, particularly for tilapia, which was the dominant migrant fish into the ponds (Teichert-Coddington et al., 1997). The pH varied from 7 to 9 indicating good conditions for biological productivity. This is advantageous for Fingerponds as it implies that new ponds may not require pre-treatment with lime. The elevated electrical conductivity was extremely high compared to Lake Victoria. While a range of 107 to 120 μScm[-1] was measured in the lake water, the conductivity in Fingerponds ranged from 4000 to 7000 μScm[-1] in 2003 and even higher in the relatively drier season of 2004. It may be speculated that mineralization in the

wetland and the subsequent increase in the ionic concentration in the wetland groundwater contributed to these high values through groundwater-pond water interaction. The high conductivity during the dry season could be the result of increased ionic concentration as the pond water volume decreased with evaporation during the dry season. The effects of high electrical conductivity on the overall pond fauna in such systems warrant investigation. In such conditions one would expect considerable energy investment by pond organisms in maintenance of the osmotic balance and this may have a negative effect on overall productivity. Tilapia is a hardy fish that tolerates extreme conditions, including brackish waters (Watanabe et al., 1985; Kamal and Mair, 2005).

Manuring the ponds with livestock manure enhanced nutrient availability for primary production as indicated by the chlorophyll *a* concentration. Nitrogen levels (NH_4-N, NO_3-N and total nitrogen) were higher in manured ponds, indicating a positive effect of fertilization with livestock manure. The increase in ammonium ion concentration in manured ponds requires caution as the equilibrium concentration with un-ionized ammonium ions is highly dependent on temperature and pH (Emerson et al., 1975). In tropical conditions, where temperature and photosynthetic activity may rise dramatically in the afternoons with the latter leading to increased pH, there is a risk of equilibrium shift from ionized ammonium to the un-ionized form (Knud-Hansen, 1998). Elevated concentrations of un-ionized ammonia ions are toxic to fish (El Shafai et al., 2004). In the present study, the observed NH_4-N concentrations in ponds that received 2500 kg ha^{-1} of manure per 2 weeks ranged from 0.8 to 1.44 mg l^{-1}. Prolonged exposure of Nile tilapia to un-ionized ammonia concentrations of 1.5 to 1.7 mg l^{-1} at temperatures of 28-33 °C can stop growth (Lin et al., 1997) while concentrations between 0.7 and 0.14 mg l^{-1} have been reported to negatively impact on fish growth (El-Shafai et al., 2004). In the Lake Victoria Fingerponds, the latter levels can be reached in the afternoons, when temperatures and pH rise above 25 °C and 8.5 respectively. This implies that there is a risk of ammonia toxicity if the ponds are heavily manured. However, during the experimental period no adverse water quality effects resulting in fish kills were observed. Ammonia not only affects fish but also zooplankton and this may also have an indirect effect on pond secondary productivity (Arauzo, 2003).

Phosphorus is often assumed to be limiting in tropical ponds (Lin et al., 1997). The soluble reactive phosphorus level was higher in manured ponds and ranged from 0.18 to 0.39 mg l^{-1} compared to a range of 0.02 to 0.06 mgl^{-1} in un-manured ponds. Similarly, the total phosphorus concentration was higher in manured ponds compared to un-manured ponds. When compared with other pond aquaculture studies, the total phosphorus concentration was within typical ranges for semi-intensive to intensive aquaculture systems (Seim et al., 1997). The DIN: DIP ratio was lower than the Redfield N:P ratio of 16:1. Using the mean concentrations of DIN and DIP and dividing these values by the corresponding Redfield number gave values of 0.05 and 0.25 for nitrogen and phosphorus respectively, indicating a possible limitation of nitrogen in the ponds.

Low to moderate manure application did not seem to impact critically on the pond environment through organic loading as BOD$_5$ remained below 15 mg l^{-1} and within the range observed in extensive and semi-intensive production (Seim et al., 1997). Semi-diurnal measurements showed early morning oxygen concentrations of about 3 mg l^{-1}. No mortalities or noticeable stress on fish were observed during the culture period. Even though manure may increase the biochemical oxygen demand, the current application rates did not result in critical dissolved oxygen concentrations especially for tilapia, which can tolerate 1 to 2 mg l^{-1} (Teichert-Coddington et al., 1997).

In pond aquaculture, light penetration through the water column is crucial for primary productivity. The proportion of the solar radiation that passes through the water column is a function of both organic (mainly algae) and inorganic (clay turbidity) particulates (Boyd, 1990). In the present study, the chlorophyll *a* concentrations ranged from 10 to 160 µg l^{-1} and were not strongly related to Secchi disk depth, SDD (Figure 7). The low SDD observed in Kusa may therefore be attributed to other factors such as clay turbidity other than algal biomass (Figure 4.6). This may have impacted negatively on pond productivity, contributing to the relatively poor productivity of the Kusa ponds compared to the Nyangera ponds.

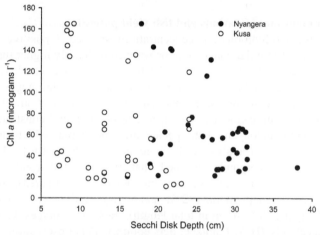

Figure 4.6: Relationship between chlorophyll *a* concentration and pond transparency (measured as Secchi disk depth) at the two experimental Fingerponds sites.

The effects of manure on pond sediment nutrients

Pond bottom sediments play a vital role in the functioning of aquaculture ponds (Seo and Boyd, 2001; Thunjai et al., 2004) and in our ponds the livestock manure application increased the nutrient status of the pond bottom sediments accordingly. The explanatory power of the regression models for total phosphorus was high, whereas that for total nitrogen was weak. The manure quality (total phosphorus and total nitrogen content) was generally good; however, while there was a noticeable effect of manuring on sediment phosphorus concentration, the effect on nitrogen concentration was unclear. There were also differences in soil nitrogen between the two sites, indicating the spatial variability and probably also the effect of manuring strategy. The relatively high sorption capacity of soils for phosphorus compared to nitrogen may partly explain the observed effect of manuring on sediment phosphorus concentration. The significant effects of time as indicated by the repeated measures ANOVA showed that there was a progressive accumulation of sediment nitrogen (although somewhat erratically) and phosphorus during the culture period due to the continuous application of manure.

Another observation was the indication of the effect of manuring strategy on the spatial distribution of nutrients in the pond sediments. Nutrients seemed to accumulate at the zone of manure application leading to differences in nitrogen and phosphorus between sediments in the deep and shallow sections of the pond. This was confirmed by the fact that in Nyangera, where manure was broadcasted daily, the distribution of pond sediment nutrients was more uniform compared to Kusa where manure was soaked in one corner and spread in the shallow end after a fortnight.

In intensive pond systems, the sediments become recipients of large amount of nutrients and chemical substances contained in food or fertilizer inputs and thus may contribute to deteriorating water quality in the subsequent culture period. The management of pond soils is therefore very important (Boyd et al., 2002). The proposed management practice for Fingerpond systems is to transfer the accumulated nutrient-rich sediments to the adjacent vegetable gardens during the dry season.

Pond environment manipulations and fish yield potential

The algal biomass (chlorophyll a concentration) showed a positive response to manuring. The chlorophyll a concentrations were considerably higher than the upper range of 71.5 μgl^{-1} recorded for Lake Victoria (Lung'ayia et al., 2000). Both chlorophyll a and fish yield biomass were comparable with the ranges observed in other extensive and semi-intensive pond aquaculture (Seim et al., 1997). Algal biomass is an important element of the aquatic food chain and constitutes the main diet for most zooplankton and fish species such as tilapia, which dominated in Fingerponds. Gut content analysis in 21 specimens of tilapia fry ($<$ 4 cm in total length) from the Fingerponds revealed that the ingested food consisted of phytoplankton (found in 81% of the individuals investigated), detritus and macrophytes (29%), zooplankton (22%) and insects and protozoa (14%). In 73 specimens of larger fish ($>$ 4 cm in total length), these percentages were 20, 52, 38 and 10%, respectively (H.A. Lamtane, pers comm.). This observation confirms the ontogenic shift from zooplankton to phytoplankton and detritus in tilapia. Pond fertilization had an indirect effect on secondary production as reflected in the increased pond water nutrient concentrations and consequently higher fish yields in manured ponds. Further enhancement of fish natural food production may be achieved by addition of periphyton substrates (Azim, 2001) or green manure (Knud-Hansen, 1998).

The fish yields are based on partial fish harvests, as complete harvest at the end of the season could not be achieved through seining. There was a high variability in yields in the manured ponds ranging from 400 to 1,000 kg ha^{-1} between the two experimental sites. This suggests a considerable spatial variability in Fingerponds fish productivity. Variable yields have been attained in integrated aquaculture systems with semi-intensive to intensive production in other projects too. Yields from conventional aquaculture ponds nearby in Kusa, stocked with tilapia and catfish and receiving a variety of inputs ranging from cattle manure, fishmeal, freshwater shrimp, slaughterhouse and brewer's wastes and household wastes, ranged from 1920 to 2285 kg ha^{-1} (Manyala, 2003). Ofori et al. (1996) reported yields of 1550 and 1940 kg ha^{-1} yr^{-1} in Ghana from ponds fertilized with cow dung and chicken manure, respectively. In Malawi, fish yields from polyculture systems ranged from 868 to 1099 kg ha^{-1} over a 150 day period (Brummett and Noble, 1995). Higher yields of 929 to 1964 kg ha^{-1} over a 126 day culture period in improved pond management systems with napier grass inputs were reported in the same region by the above authors. Fish yields of 3000 to 5000 kg ha^{-1} yr^{-1} per hectare per year have been reported in Collaborative Research Support Programme (CRSP) experimental sites (Lin et al., 1997). However it must be noted that these were mainly on-station and highly managed systems with selected fish species and optimized stocking densities. In such systems, pond fertilization is often achieved by both inorganic and organic nutrient inputs. In commercial systems where higher yields and large fish are desired, supplemental feeds are added. Fingerponds are similar to capture based aquaculture (CBA), whereby "wild seeds" are obtained and cultured to marketable size (FAO, 2004 a). Such production systems are intermediate between capture fishery and true aquaculture. Thus, although the yields from Fingerponds are lower than conventional aquaculture ponds they represent a considerable improvement when compared to the natural capture fishery from floodplains, lakes and reservoirs (Table 4.6). With more experience, improved harvesting technology and adaptive management, yields may be expected to increase.

Table 4.6: Fisheries production of selected African wetlands and inland water bodies.

Ecosystem	System	Fish yield (kg ha^{-1})	Source
Floodplain	Niger system	31.2-49.6	(Cited in Dugan, 2003) Modified from Lévêque,1999; Welcomme, 1989
	Kafue flats	15.6	
	Senegal system	54.7	
	Nile Sudd	8.8	"
Reservoirs	Lake Nasser	6-25	"
	Kariba Dam	30-40	"
	Kainji Dam	35-47	"
	Lagdo	175-300	"
Lakes	Lake Baringo	10-50	"
	Lake Naivasha	5-60	"
	Lake Malawi	35-45	"
	Lake Tanganyika	90	"
	Lake Victoria	29-59	"
Wetlands	Hadejia-Nguru (Nigeria)	49	Ita, 1993
	Ogun and Oshun (Nigeria)	40	Ita, 1993

In spite of the promising potential of Fingerponds in the enhancement of wetland fishery production, some challenges remain. One of the main challenges encountered was the early spawning and excessive recruitment by tilapia, which led to overstocking. A similar observation was made in experimental Fingerponds in Uganda (R.C. Kaggwa, pers. comm.). It appears therefore that this may be a common problem in Fingerponds. Overstocking has a negative effect on the overall fish yields due to induced competition and stress (Glasser and Oswald, 2001). An attempt to separate the males and females in Kusa by hand sexing in 2004 gave some positive results although it was not a complete success. Excessive fingerling production can also be controlled effectively by stocking the ponds with large *Clarias gariepinus* (6-130 grams) at rate of 4000 to 10000 per hectare (de Graaf, 1996). The natural stocking densities of *Clarias gariepinus* in the experimental ponds ranged from 365 to 1614 individuals per hectare with a weight range of 2 to

126 g and a modal frequency of 12 g. This Clariid density was inadequate for effective cropping of tilapia.

The desire for very large-sized tilapia (>500g) is not a priority in Fingerponds. In rural areas, palm size tilapia (150-250 g) is generally acceptable for household consumption. Although the small-sized fish dominating these production systems may not be attractive for local markets, they can contribute significantly to household per capita protein supply (Chapter 6). Continuous partial harvesting may spread fish supply to the households over a longer period compared to the short-lived traditional wetland capture fishery. During the fish census and intra-seasonal harvests, the excess fingerlings can be transferred to under-stocked ponds or the adjacent natural water bodies.

Another challenge is how to control predation by otters, monitor lizards and fish-eating birds. The fact that the ponds are located far away from the farmer's house increases the chances of theft. However, the local communities have been cultivating crops in the wetlands with limited cases of theft. The societal norms and existing local rules usually enforced by the village government and clan elders are expected to apply also to Fingerponds.

Conclusion

Fingerponds have the potential for self-stocking with different fish species which can utilize the diverse niches available in the ponds. Fertilizing the ponds with livestock manure had a positive effect on fish growth and yields. However, there was an indication of nitrogen limitation in the ponds. Nevertheless, with intermediate level of management such as the addition of livestock manure, fish yields of 500 to 1000 kg ha^{-1} could be achieved. Higher yields can be achieved with improved pond management. There is a tendency of over-stocking during the culture period due to high recruitment by the tilapia, which is the dominant fish species in the Lake Victoria Fingerponds.

In such ponds with un-regulated water supply, care should be taken not to overload the ponds with organic manure as this could lead to the deterioration of water quality, especially to ammonia toxicity, and subsequent negative effects on pond fauna. During the subsequent culture periods, manure application rate should be adaptive so that, as the water level declines towards the dry season, a good water quality is maintained.

Acknowledgements

I wish to acknowledge the financial support from the European Union Fingerponds project Contract no. ICA4-CT-2001-10037. Additional funding for fieldwork was provided by the International Foundation for Science, Stockholm, Sweden and Swedish International Development Cooperation Agency Department for Natural Resources and the Environment (Sida NATUR), STOCKHOLM, Sweden, through grant no. W/3427-1. I appreciate the assistance from the Fingerponds Project research team and the local communities at the respective study sites. I would also like to express my gratitude to Mr. Julius Manyala of Moi University Fisheries Department for assisting in fish identification.

References

Aguilar-Manjarez, J., Nath, S.S., 1998. A strategic re-assessment of fish farming potential in Africa. CIFA Technical Paper. No. 32. Food and Agriculture Organization of the United Nations, (FAO) Rome, 170 pp.

APHA, 1992. Standard methods for the examination of water and wastewater. 18th edition. American Public Health Association, Washington DC, United States of America.

APHA, 1995. Standard methods for the examination of water and wastewater.19th edition. American Public Health Association, Washington DC, United States of America.

Arauzo, M., 2003. Harmful effects of un-ionized ammonia on the zooplankton community of a deep wastewater treatment pond. Water Research 37 (5), 1048-1054.

Azim, M.E., 2001. The potential of periphyton-based aquaculture production systems. PhD Thesis, Wageningen University, 219 pp.

Boyd C, E., Bowman J. R., 1997. Pond bottom soils. In: Egna, H.S., Boyd, C.E. (eds.), Dynamics of Pond Aquaculture, CRC press, Boca Raton, Florida, pp. 135-162.

Boyd, C.E., 1990. Water quality in ponds for aquaculture. Alabama Agricultural Experimental Station, Auburn University, Auburn, Alabama.

Boyd, C.E., Wood, C.W., Thunjai, T., 2002. Aquaculture pond bottom soil quality management. Aquaculture Collaborative Research Support Program, Oregon State University, Corvallis, Oregon, 41 pp.

Brummett, R.E., 1999. Integrated aquaculture in sub-Saharan Africa. Environment Development and Sustainability 1, 315-321.

Brummett, R.E., Noble, R., 1995. Aquaculture for African smallholders. ICLARM Technical Report 46, 69 pp.

COFAD, 2002. Back to basics: Traditional inland fisheries management and enhancement systems in sub-Saharan Africa and their potential for development. Deutche Gesselschaft fur Technische Zussamenarbeit (GTZ) GmbH, Eschborn, 203 pp.

De Graaf, G., Galemoni, F., Banzoussi, B., 1996. Recruitment control of Nile Tilapia by the catfish *Clarias gariepinus* (Burchell, 1822) and African Snakehead *Ophiocephalus obscuris*. I. A biological analysis. Aquaculture 146, 85-100.

Delincé, G., 2000. The ecology of the fish pond ecosystem with special reference to Africa. Kluwer Academic Publishers, Dordrecht.

Diana, J.S., Lin, C.K., Schneeberger, P.J., 1991. Relationships among nutrient inputs, water nutrient concentrations, primary production and yield of *Oreochromis niloticus* in ponds. Aquaculture 92, 323-341.

Dugan, P., 2003. Investing in Africa: the Worldfish Center's African strategy in summary. In: M.J Williams (ed.) NAGA World Fish Center Quarterly vol. 26 No. 3, pp. 4-7.

El-Shafai, S.A., El-Gohary, F.A., Nasr, F.A., van der Steen, N.P., Gijzen, H.J., 2004. Chronic ammonia toxicity to duckweed-fed tilapia (*Oreochromis niliticus*). Aquaculture 232 (1-4), 117-127.

Emerson, K.R.C, Russo, R.E., Thurston, R.V., 1975. Aqueous ammonia equilibrium calculation: effect of pH and temperature. Journal of the Fisheries Research Board of Canada 32, 2379-2383.

FAO, 2004 b. Aquaculture extension in sub-Saharan Africa. Fisheries Circular No. 1002. Food and Agriculture Organization of the United Nations, Rome 55 pp.

FAO, 2004 a. State of World Fisheries and Aquaculture (SOFIA). Food and Agriculture Organization of the United Nations, Fisheries Department, Rome, 153 pp.

FAO, 2005. FAOSTAT data, http://faostat.fao.org/ accessed February 2005.

Glasser, F., Oswald, M., 2001. High stocking densities reduce *Oreochromis niloticus* yields: model building to aid optimization of production. Aquatic Living Resources, 14, 319-326.

Halwart, M., van Dam, A.A. (eds.)., 2006. Integrated irrigation and aquaculture in West Africa: concepts, practices and potential. Food and Agriculture Organization of the United Nations, Rome, 181 pp.

Ita, E.O., 1993. Inland fishery resources of Nigeria. CIFA Occasional Papers No. 20, Food and Agriculture Organisation of the United Nations (FAO) Rome, 120 pp.

Jamu, D.M., Ayinla, O.A., 2003. Potential for the development of aquaculture in Africa. In: M.J. Williams (ed.) NAGA World Fish Center Quarterly vol. 26 No. 3, pp. 9-13.

Kamal, A.H.M.M., Mair, G.C., 2005. Salinity tolerance in superior genotypes of *Oreochromis niliticus*, *Oreochromis mossambicus* and their hybrids. Aquaculture 247, 189-201.

Knud-Hansen, F.C., 1998. Pond fertilization: ecological approach and practical application. Pond Dynamics / Aquaculture Collaborative Research Support Programme, Oregon State University, Corvallis OR, 125 pp.

Lévêque, C., 1999. Les poisons dex eaux continentals africaines: diversité, écologie, utilisation par l'homme. IRD, Paris, 521 pp.

Li, W., Yang, Q., 1998. Wetland utilization in Lake Taihu for fish farming and improvement of lake water quality. Ecological Engineering 5, 107-121.

Lin, C.K., Teichert-Coddington, R., Green, B.W., Veverica, K.L. 1997. Fertilization regimes. In: Egna, H.S., Boyd, C.E. (eds.), Dynamics of pond Aquaculture, CRC press, Boca Raton, Florida, pp.73-107

Lind, O.T., 1979. Handbook of common methods in limnology. C.V. Mosby, St. Louis MO. 199 pp.

Lungáyia, H.B.O, M'Harzi, A., Tackx, M., Gichuki, J., Symoens, J. J., 2000. Phytoplankton community structure and environment in the Kenyan waters of Lake Victoria. Freshwater Biology 43, 529-543.

Machena, C., Moehl, J., 2001. Sub-Saharan African aquaculture: regional summary. In: R.P. Subasinghe, P. Buana, M.J. Phillips, C. Hough, S.E. McGladdery and J.R Arthur (eds.). Aquaculture in the Third Millennium. Technical Proceedings of the conference on aquaculture in the Third Millennium, Bangkok, Thailand, 20-25 February, 2000, NACA, Bangkok and FAO, Rome, pp. 341-355.

Manyala, J.O., 2003. Fish farming report. Consultancy report for Regional Land Management Unit (Relma) Kusa community Development project (KCDP).

Odada, E. O., Olago, D. O., Kulindwa, K., Ntiba, M., Wandiga, S., 2004 Mitigation of environmental problems in Lake Victoria, East Africa: causal chain and policy options analyses. Ambio 33, No. 1-2 pp. 13-23.

Ofori, J., Aban, E.K., Otoo, E., Wakatsuki, T., 2005. Rice-fish culture: an option for smallholder Sawah rice farmers of West African lowlands. Ecological Engineering 24, 235-241.

Ofori, J.K., Asamoah, A., Prein, M., 1996. Experiments for integrated agriculture aquaculture system design. In: M. Prein, J.K. Ofori and C. Lightfoot (eds.) Research for the future development of aquaculture in Ghana, ICLARM Conf. Proc. No. 42.

Okalebo, J.R., Gathua, K.W., Woomer, P.L., 2002. Laboratory methods of soil and plant analysis; A working manual. 2nd edition, SACRED Africa, Nairobi, 128 pp.

Pechar, L., 1987. Use of acetone:methanol mixture for the determination of extraction and spectrometric of chlorophyll *a* in phytoplankton. Arch. Hydrobiol. Suppl. 78, 99-117.

Prein, P., Ruddle, K., Ofori, J.K., Lightfoot, C., 1995. Assessment of integrated aquaculture potential using a farmer-participatory approach: a case study in Ghana. ICLARM Technical Report 42, 90 p.

Seim, W.K., Boyd, C.E., Diana, S.J., 1997. Environmental considerations. In: Egna, H.S., Boyd, C.E. (eds.), Dynamics of pond Aquaculture, CRC press, Boca Raton, Florida, pp. 163-182.

Seo, J., Boyd, C.E., 2001. Effects of bottom soil management practices on water quality improvement in channel catfish *Ictalurus punctutus* ponds. Aquacultural Engineering 25, 83-97.

Teichert-Coddington, D.R., Popma, T.J., Lovshin, L.L., 1997. Attributes of tropical pond-cultured fish. In: Egna, H.S., Boyd, C.E. (eds.), Dynamics of Pond Aquaculture, CRC press, Boca Raton, Florida, pp.183-198.

Thunjai, T., Boyd, C.E., Boonyaratpalin, M., 2004. Bottom soils quality in tilapia ponds of different age in Thailand. Aquaculture Research 35, 698-705.

Van Dam, A.A., 1990. Multiple regression analysis of accumulated data from aquaculture: a rice-fish culture example. Aquaculture and Fisheries Management 21, 1-15.

Van Deventer, J.S., Platts, W.S., 1985. A computer software system for entering, managing and analyzing fish capture data from streams. USDA forest service research note INT-352, Intermountain research station, Ogden, Utah, 12 pp.

Van Oijen, M.J.P. (1995). Appendix I. Key to Lake Victoria fishes other than haplochromine cichlids. In: F. Witte and W.L.T. van Densen (eds.) Fish stocks and fisheries of Lake Victoria. A handbook for field observations. Samara Publishing Limited, Dyfed, Great Britain, pp. 209-300.

Veverica, K.L., Bowman, J., Popma, T., 2001. Global experiment: Optimisation of nitrogen fertilization rate in freshwater tilapia production ponds. In: A. Gupta, K. McElwee, D. Burke, J. Burright, X. Cummings and H. Egna (eds.). Eighteenth Annual Technical Report. Pond Dynamics/Aquaculture CRSP, Oregon State University, Covallis, Oregon, pp.13-22.

Wahby, S.D., 1974. Fertilising fishponds I- Chemistry of the waters. Aquaculture 3, 245-259.

Watanabe, W.O., Kuo, C.-M., Huang, M.-C., 1985. Salinity tolerance of tilapias *Oreochromis aureus*, *O. niloticus,* and an *O. mossambicus* × *O. niloticus* hybrid. ICLARM Technical Report 16, 22 pp.

Welcomme, R. 1989. Review of the present state of knowledge of fish stocks and fisheries of Africa rivers. In: D.P Dodge (ed.) Proceedings of the International large Rivers Symposium, Canadian Special Publication of Fisheries and Aquatic Science 106, 515-532.

Wohlfarth, G.W., Schroeder, G.L. 1979. Use of manure in fish farming-A review. Agricultural Wastes 1, 279-299.

Yussoff, F.M., McNabb, C.D., 1989. Effects of nutrient availability on primary productivity and fish production in fertilized ponds. Aquaculture 78, 303-319.

Chapter 5

A quantitative assessment of nutrient flows in an integrated agriculture-aquaculture farming system at the shores of Lake Victoria, Kenya

Abstract

The rural farming systems around Lake Victoria are predominantly subsistence with integrated crop and livestock production. The results presented here are based on integrated experimental smallholder aquaculture systems (Fingerponds) at the Lake Victoria wetlands. The overall objective was to increase fish protein supply to the households and the diversity in farming activities. The rural farming system was characterized using natural transect mapping alongside identification of bioresource flows between the system components. Nutrient flows were analyzed using Ecopath with Ecosim 5.1 software using nitrogen as the model currency. The model result scenario with and without the wetland demonstrated the importance of the natural wetland in the overall agroecosystem nutrient flows. The farming system is characterized by low nutrient throughput associated with low productivity. Nutrient balance at the Fingerponds sub-system level was highly positive compared to maize production, which is the dominant activity in the terrestrial ecosystem. Diversification of the farming system through integration of Fingerponds increases the nutrient flow pathways and functional diversity. However, Fingerponds had minimal impact on the agroecosystem performance indicators such as Biomass to throughput (B/E) and production to biomass (P/B) ratios, which are usually used to gauge ecosystem maturity and hence sustainability potential. This is probably because the overall farming systems productivity is low and Fingerponds is a small component of a larger agroecosystem. Nevertheless, modelling such systems with Ecopath provided a better insight to the agroecosystem nutrient flows.

Key words: Lake Victoria-Kenya, integrated farming systems, wetlands, Fingerponds, ECOPATH, sustainability

Introduction

In sub-Saharan Africa, smallholder rural households are faced with increasing vulnerability to food shortages. Forced by unreliable weather and frequent failure of rain-fed upland crop production, the local communities around wetlands in the Lake Victoria basin have increasingly turned the wetland margin into crop production patches. The rapidly growing human population is expected to have a profound impact on the wetland ecosystems (Kairu, 2001). There is a need for production systems that promote a balance between human needs and ecosystems conservation (Salafsky and Wollenberg, 2000).

Despite considerable efforts to develop appropriate farming system technologies in many parts of the world, most farming systems in sub-Saharan Africa are still characterized by high negative nutrient balances (De Jager et al., 2001; Roy et al., 2003). In recent years, Asian integrated farming systems have been recognized as examples of sustainable natural resource management (Haylor and Bhutta, 1997; Fernando and Halwart, 2000; FAO, 2001; Prein 2002). One of the benefits of integration lies in synergy between system components through nutrient recycling and more efficient production (i.e. more production output per unit of nutrient input). Efforts to describe the ecology of integrated agro-ecosystems and improve the understanding of their functioning have led to the emergence of agricultural ecology or agro-ecology as a discipline (Conway, 1987; Tivy, 1990; Altieri, 2002;). An agroecosystem *sensu lato* refers to an ecological system in which the farmer plays a key role in influencing the functionality of its components. In this paper the term agro-ecosystem is used to refer to an integrated rural farming system in Kenya with experimental smallholder aquaculture-agriculture (Fingerponds) in a floodplain wetland.

Fingerponds are earthen ponds excavated in fringe wetlands during the dry season. The soil removed is spread to create raised beds for vegetable production. The ponds resemble natural flood pools used for wetland fish capture by local communities while the gardens are a continuation of the normally-existing seasonal swamp margin vegetable patches. When several of these narrow channel-like ponds and the adjacent gardens are constructed close to one another at the lake/swamp or swamp/land interface, they appear like " fingers" into the emergent macrophyte zone from a bird's eye view. This is why they are called "Finger ponds". The ponds are stocked by wild fish during annual flooding of the wetland with fish culture and garden management starting after flood recession. Manure from livestock and vegetable wastes is applied to the ponds to stimulate the production of natural fish food while water from the ponds may be used for irrigation. The use of livestock manure for pond manuring creates a link between terrestrial and wetland farming systems and forms the basis for integration of the ponds into the wetland farming systems. Initial studies in Kenya have shown that a 400 m^2 Fingerpond can produce about 8-20 kg of fish during a 5-6 month culture period (Kipkemboi et al., 2006).

Wetland agroecosystems in Kenya may be economically and socially important because they contribute to food production and income generation and therefore to the livelihoods of wetland communities. In addition, the natural wetland biomass dominated by *Cyperus papyrus*, *Phragmites* sp. and *Typha domingensis* and other sedges are important for biomass harvesting for various uses (Gichuki et al., 2001). Currently, expansion of wetland agriculture is not ecologically sustainable because large areas of papyrus wetland are converted to seasonal crop production, leading effectively to destruction of papyrus swamps and their natural functions. In addition, agricultural development may lead to the import of nutrients into wetlands in the form of chemical fertilizers, leading to eutrophication in the surrounding water bodies. In order to be ecologically sustainable, integration of Fingerponds into the existing farming system should increase food production from the same area of land whilst respecting the ecological functioning of the wetland. This may prevent more encroachment on the wetlands and also contribute to more efficient use of nutrients in the whole system (thus preventing discharge of nutrients into the environment).

Like natural ecosystems, agroecosystems consist of components that interact to give characteristic flows of energy and matter. These components create an ecological complexity, which needs to be unraveled in order to be understood and managed sustainably. Systems analysis can provide insights into the flow characteristics and performance of such integrated systems and form the basis for an overall sustainable management approach (Odum, 1983, Dalsgaard, 1995, 1997; Dalsgaard and Christensen, 1997; Dalsgaard and Oficial, 1998; Liang, 1998). The objectives of this study were to: (1) characterize and quantify the bioresource flows and efficiency within the current wetland farming system using a systems analysis approach; and (2) evaluate the effects of integration of Fingerponds on productivity and nutrient efficiency of the agro-ecological system.

Methods

The study site

Fingerponds were established in Kayano village in Kusa, Kenya in 2002. The village lies adjacent to the river Nyando floodplain wetland in the Lake Victoria catchment of Kenya (Lower Nyakach division, Nyando District, Nyanza province at 0° 18' 1.2 "N and 34° 53' 21.3"E). The farming systems are predominantly rain-fed mixed crop systems integrated with livestock production in the terrestrial parts and seasonal crop cultivation in the wetland. Rainfall is variable with mean annual rainfall between 750 and 1000 mm. The rainfall pattern determines the calendar of farming activities types during the year (Figure 5.1). Normally the rain is biannual with the wettest period in April-May. However, over the recent years rainfall has become erratic leading to frequent crop failures in the rain-fed agriculture (personal observation).

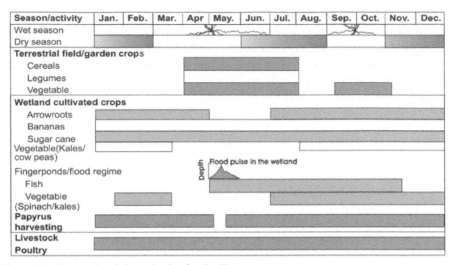

Figure 5.1: Seasonal activity calendar for the Kusa agroecosystem

Modelling nutrient flows in integrated farming systems

Ecopath mass balance modelling was used to provide insights into the agro-ecosystem attributes. Ecopath was originally developed for aquatic ecosystems analysis by Polovina (1984) and modified by Christensen and Pauly (1992) and Pauly et al. (2000). Ecopath is based on a static mass balance approach and unlike dynamic modelling, it provides snapshot information about the system for a specific period of time. To construct an Ecopath model, the ecosystem is divided into a number of functional groups, each group distinguished by its function and/or life form. Functional groups can be a single species or guilds of species, each described by the following master equation:

Production = Catches + Predation mortality + Biomass accumulation + Net migration + Other mortality   (5.1)

Aquatic and terrestrial ecosystems are governed by the same principles of mass and energy conservation; hence Ecopath can also be applied to agroecosystems (van Dam et al., 1993; Ruddle and Christensen, 1993; Lightfoot et al., 1993; Dalsgaard and Christensen, 1997; Dalsgaard and Oficial, 1998; Dalsgaard and Prein, 1999), in which functional groups are often crops or animals coinciding with farm enterprises, the above relationship has been adapted as follows:

Production = Harvest + Resource flows + Biomass accumulation + (Losses - Import) + Flow to detritus   (5.2)

This relationship can be expressed as:

$$P = H + RF + BA + (L - IM) + OM \tag{5.3}$$

where Production (P) refers to generation of new tissue/biomass, Harvest (H) refers to the biomass extracted from the group, e.g., as products sold to the market, bartered or given away as gifts; resource flows (RF) refers to the biomass moved to other agro-ecosystem groups; losses (L) often occur through natural processes (e.g., volatilization or leaching of nutrients); and imports (IM) occur through fertilizers and feeds purchased from the market as well as gains through natural processes. Flow to detritus or other mortality (OM) is production that is not harvested or consumed within the system and by Ecopath default is returned to the detritus box. The above relationship can be re-expressed as

$$B_i.(P/B)_i EE_i - H_i + IM_i - \sum_{j=1}^{n} B_j.(Q/B)_j.RF_{ij} - BA_i - EX_i = 0 \quad (5.4)$$

where B_i is the biomass of group i, (P/B_i) is the production biomass ratio of group i, EE is the ecotrophic efficiency the fraction of production that is consumed within the system, Q/B_i is the consumption/biomass ratio of group i, H_i is harvest, IM_i are imports to group I from outside in form of fertilizers, manure or organic residues, BA_i is biomass accumulation, RF_{ij} is the fraction of resource flow fraction of i to j and EX_i is export from group i. For each functional group in the system, Ecopath uses equation 5.4 to define the mass balance relationship between consumption, production and net export for a given period of time (usually one year). Ecopath sets up as many equations as there are functional groups in the system and solves the system of linear equations for the missing parameter for each group. Of the four parameters (B, P/B, Q/B and EE), three must be provided. Additionally, H_i and RF_{ij} must be defined. Data collection was geared at finding parameter values for each functional group in the system.

Model conceptualization and compartmentalization

First, a general inventory of farming activities and natural resources was made and a seasonal calendar and resource transect were constructed from information obtained through farm walks and informal interviews. The study focused on major functional groups based on function and life forms, which form distinctive management entities. The farming system includes two distinct subsystems: terrestrial and wetland (Figure 5.2). The terrestrial system was dominated by terrestrial field and garden crops, multipurpose trees and shrubs. The wetland sub-system consisted of seasonal crops: arrowroots, vegetables, sugar cane and bananas. Another important wetland-based household activity was papyrus biomass harvesting mainly for mat-making. The components associated with Fingerpond technology were fish, vegetable and phytoplankton. Households played a big role in the agroecosystem bioresource flows in this study, hence the inclusion of 12 households involved in the Fingerponds pilot study in the model.

Soil is the source of nutrients that determines the productivity of both terrestrial and wetland ecosystems. Soil types were broadly categorized into two types: the terrestrial and wetland soils. In the model the wetland soil was further sub-divided into three categories; papyrus area soil with little human interference, wetland cultivated area (including the Fingerpond gardens) and pond soil (Fingerponds).

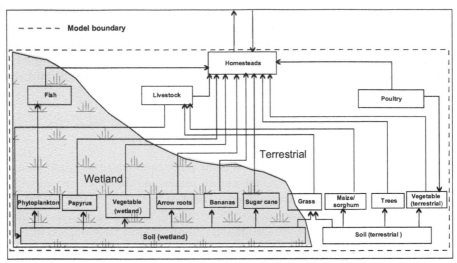

Figure 5.2: A conceptual model representation showing the wetland-terrestrial ecosystem linkage

Due to practical constraints and the level of analysis for this study, organisms which are likely to be present in the system but are not utilized directly by the household were not included in the model. These included reptiles, amphibians, several mammals such as hippopotami, otters, rodents and other wetland wildlife that may cause exports from (e.g., through predation), but also imports into the system (e.g., through manure) but their impacts were thought to be small. Other organisms that were excluded from the model were micro-organisms, zooplankton, insects and other macroinvertebrates whose biomasses were considered to be small in relation to other important groups. Some of the organisms in this group are pests or cause diseases and their effects on the productivity of the various farming systems components is assumed to be captured in the net yields.

Data collection, analysis and model calibration
Data collected aimed at obtaining the key input parameters for Ecopath models; B, Q/B, P/B and EE. Parameters values were obtained through two main approaches: field data collection and secondary information. Field data collection and processing was based on the approach used by Dalsgaard and Oficial (1998), Lightfoot et al. (2000), Schlaman (2003) and Luoga (2005). Resource flows, biomasses and harvests for each farm enterprise/crop were monitored for 12 household farms for ten months from May 2004 to February 2005. Pre-tested semi-structured questionnaires were used to collect information from the households on a monthly basis. As rural households do not keep records of the items consumed, the questionnaire focused on the last 7 days before the time of the interview, for which the respondents mainly women, provided estimates of household consumption and other farming system resource flows. On many occasions the estimates were in local measurement units such as the number of *gorogoros* (approximately 2 kg tin of grains), number of bottles of milk, etc. All measurements were verified through actual weighing of samples and converted to the International System of units (SI), in this case, kilograms. Additionally, direct field measurements on various farming systems components biomass production were carried out prior to arrangement with the

households. Fingerponds fish yields were based on an average biomass of 20.5 kg fish per 192 m^2 pond size obtained from the experimental Fingerponds. Vegetable yields were obtained from records of spinach harvests from Kusa site of an average of 85 kg per season per 192 m^2 Fingerpond garden.

Nitrogen was identified as the model currency and was expressed as mg N m^{-2} yr^{-1}. Nitrogen is one of the "big three" NPK (nitrogen, phosphorus and potassium) elements often considered as the key nutrients in agricultural production. Nitrogen and phosphorus are frequently identified as limiting both in tropical and temperate agricultural production systems. Nitrogen is also important as the key element of food protein, which is an important factor in household food security. Data collected from the 12 households was used for the model input. To convert kg biomass to mg N m^{-2}yr^{-1}, the dry matter and nitrogen content of various groups were obtained from literature sources (Leung, 1968; Zamora and Baguo, 1984; Garrow et al., 1993; Ntiamoa-Baidu, 1998; http://weather.nmsu.edu/ hydrology/wastewater/plant-nitrogen-content.htm). The information obtained above was used to generate a diet composition matrix, which defines the interaction between the various groups in the system (Table 5.1).

The overall modelling approach was centered on the introduction of fingerponds within the existing farming system activities. To expand the model analysis horizon, five scenarios were used. Scenario 0 assumed a without Fingerpond status in the farming system and therefore excludes Fingerponds from the model. Scenario 1 assumed one Fingerpond (approximately 200 m^2 each for both pond and an adjacent vegetable garden) per household. This would be a more realistic scenario compared to an experimental situation of joint ownership of 4 Fingerponds by 12 households involved in the pilot study. Scenario 2 assumed an increased ownership of Fingerponds to 2 per household while Scenario 3 assumes four Fingerponds per household. In Scenario 4 the wetland components were excluded from the model in order to evaluate the importance of the natural wetland in the overall farming system nutrient flow network. Scenario 5 is similar to scenario 1, except that the household has been excluded from the model. All assumptions are based on the layout of experimental Fingerponds in Kenya.

Results

The agroecosystem bioresource flow characteristics

Figure 5.3 shows the agroecosystem components and resource transect as revealed from the survey. This conceptual representation also shows the links between various farming system components across terrestrial and wetland ecosystems. The household is the central part of the farming system as all resources flow into or through it. The terrestrial components are characterized by cultivated crops, livestock and trees while the wetland activities comprise natural biomass harvesting and seasonal crop cultivation. The terrestrial field crops consist of cereals (maize, *Zea mays* and sorghum, *Sorghum* spp.) and legumes (common bean, *Phaseolus vulgaris* and groundnuts, *Arachis hypogea*) while vegetable (kales, *Brassica oleracea* and cowpeas, *Vigna unguiculata*) and fruit crops, mainly bananas (*Musa* spp.) and pawpaw (*Carica papaya*), are grown in the kitchen garden near the homestead. The multipurpose trees are dominated by *Grevillea robusta, Euphorbia sp., Cassia siamea* grown as edge trees around the compound and boundary of the household land. Animal production is composed of ruminants (predominantly the East Africa zebu, *Bos indicus,* sheep, O*vis* sp. and goats, *Capra* sp.); and poultry (local chicken, *Gallus domesticus*). The wetland production systems consist of seasonal crop cultivation of arrowroots (*Dioscorea* sp.), bananas (*Musa* spp.), sugar cane (*Saccharrum officinarum)*, vegetables (predominantly kales, cowpeas and tomatoes, *Lycoperscicon esculentum)*. Natural wetland products harvested are wild fish capture and papyrus (*Cyperus papyrus*), which, although not directly consumed by the household, is harvested for mat-making (for sale) besides other uses (Kipkemboi et al., 2006). Currently, traditional post-flood seasonal wild fish capture in the wetlands is not important to household food supply and has been ignored in the model (Chapter 6). Ruminants frequently traverse into the wetland as livestock often graze at the wetland margin. In terms of "harvests", papyrus constitutes the largest fraction followed by arrowroots and fruits (mainly bananas) (Figure 5.4).

Fingerponds form an additional component added to the integrated crop-livestock cum wetland biomass harvest system. The Fingerpond components considered in this study were: fish (predominantly *Oreochromis* spp.), raised bed vegetables (spinach, *Spinacia oleracea*) and phytoplankton. Fingerpond systems create three new links within the farming system: one within the ponds (fish and pond natural food interaction); the second is between the ponds and the household (household fish consumption); and the third is with the livestock component (pond manuring). Consequently, the number of bioresource flow increases by three with two potential recycling pathways.

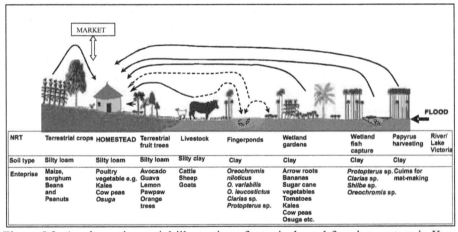

Figure 5.3: A schematic spatial illustration of a typical rural farming system in Kusa depicting the natural resource transect, flows and diversity in components (dashed flows indicate new flows associated with Fingerponds).

Table 5.1: Resource flow matrix (diet composition)

Prey \ Predator	1	2	3	4	5	6	7	8	9	10	11	12	13	14	15	16	17
1 Household																	
2 Poultry	0.0004																
3 Fish FP	0.079																
4 Ruminants	0.004																
5 Multipurp.trees	0.070			0.002													
6 Terrest. veg.	0.010	0.01															
7 Phytoplankton FP			0.30														
8 Sugar cane	0.003																
9 Fruit trees	0.018																
10 Wetland veg.	0.109																
11 Sweet potatoes	0.001																
12 Arrowroots	0.054																
13 Cereal crops	0.300	0.01		0.010													
14 Legumes	0.056	0.02															
15 Grass		0.46		0.850													
16 Papyrus							1.00										
17 Veg. FP	0.016																
18 Detritus 1					1.00	1.00										1.0	
19 Detritus 4		0.29							0.25		1.0		1.0	1.0	0.9		
20 Detritus 2			0.70						0.75	1.0		1.0					
21 Detritus 3								1.00							0.1		1.0
Import	0.281	0.21		0.140													
Sum	1.000	1.00	1.00	1.000	1.00	1.00	1.00	1.00	1.00	1.0	1.0	1.0	1.0	1.0	1.0	1.0	1.0

FP is Fingerponds, Multipurp. is multipurpose, Terrest. is terrestrial, Veg. is vegetable, Detritus 1,2,3 and 4 are wetland (papyrus zone), pond (Fingerpond), wetland (seasonal gardens) and terrestrial soils respectively

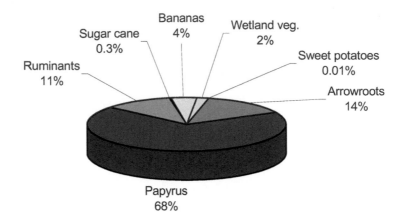

Figure 5.4: Percentage composition of harvests from the Kusa integrated farming system (based on mg N m^{-2} yr^{-1} harvested at farm level)

Model results and validation

Three basic inputs parameters (B, P/B, and Q/B) were entered on the basic inputs screen. Through basic parametization, Ecopath estimated the fourth item of the master equation, the ecotrophic efficiency (EE). Table 5.2 shows the model basic estimates computed for various groups. At the first run of the model, the estimated EE for cereals was 1.16. This implied that the model was unbalanced (Christensen et al., 2004). To balance the model, P/B, Q/B and biomass estimates were checked for reliability. Since these are seasonal/annual crops and follow a sigmoid growth curve, hence the assumption that the value of average standing biomass is roughly half of the total plant growth, the P/B and Q/B ratios were considered reliable. The diet composition at consumer level was then checked and adjusted to obtain a balanced model.

The ecotrophic efficiency (EE) values of the balanced model were less than 1 and agreed with the suggestions by Christensen and Pauly (1992) and Dalsgaard and Oficial (1998). The estimated EE value for all detritus was less than 1 indicating accumulation of nitrogen. The ecotrophic efficiency (EE) values of the balanced model were less than 1 and agreed with the suggestions by Christensen and Pauly (1992) and Dalsgaard and Oficial (1998). The estimated EE value for all detritus was less than 1 indicating accumulation of nitrogen. The EE value of the pond soil was comparatively lower than that of terrestrial and wetland cultivated area, indicating a relatively higher accumulation of nutrient, probably due to addition of livestock ruminant manure into the ponds. The ecotrophic efficiency of detritus is an important diagnostic feature for the status of the agroecosystem resource base. For instance, the EE value for the wetland papyrus zone and terrestrial soils approached a steady state and it can be inferred that there is high nutrient removal from the system components. Another important feature of this farming system is the low utilization of ruminants and multipurpose trees reflected by low EE values.

Table 5.2: Ecopath model basic parameter output results based on scenario 1

No	Group name	TL	HA	B (mg N/m$^-$)	P/B yr^{-1}	Q/B yr^{-1}	EE	H
1	Household	2.96	1.000	292.65	0.01	1.02	0	
2	Poultry	2.62	0.046	17.03	0.71	8.89	0.01	
3	Fish FP	2.15	0.007	14.15	1.51	15.00	0.927	
4	Ruminants	3.00	0.285	1811.46	0.567	1.347	0.119	120.78
5	Multipurp. trees	1.75	0.037	426.44	0.56	0.56	0.138	
6	Terrest. veg.	2.00	0.017	2.68	2.00	2.00	0.765	
7	Phytoplank. FP	1.50	0.007	24.47	24.00	2.00	0.108	
8	Sugar cane	1.90	0.040	6.55	2.00	2.00	0.345	3.56
9	Fruit trees	2.00	0.055	80.28	0.78	0.78	0.823	44.92
10	Wetland veg.	2.00	0.039	68.59	2.00	2.00	0.39	20.99
11	Sweet potatoes	2.00	0.004	0.52	2.00	2.00	0.456	0.14
12	Arrowroots	2.00	0.059	89.65	2.00	2.00	0.965	157.00
13	Cereal crops	2.00	0.241	55.76	2.00	2.00	0.931	
14	Legumes	1.50	0.099	13.27	2.00	2.00	0.739	
15	Grass	2.00	0.390	561.52	4.00	4.00	0.954	
16	Papyrus	2.00	0.021	750.84	2.00	2.00	0.497	746.03
17	Veg. FP	2.00	0.009	5.22	2.00	2.00	0.449	
18	Detritus 1	1.00	0.021	41290.2			0.903	
19	Detritus 4	1.00	0.805	391345.6			0.869	
20	Detritus 2	1.00	0.009	3628.80			0.577	
21	Detritus 3	1.00	0.183	141352.4			0.701	

TL is the trophic level, HA is the habitat area, B, P/B, Q/B, EE and H remain as defined earlier in section 2.3. Detritus 1,2,3 and 4 are wetland (papyrus zone), pond (Fingerpond), wetland (seasonal gardens) and terrestrial soils respectively

For most rural households around Lake Victoria, livestock are kept as a form of saving/insurance for emergency cash requirements and are rarely consumed or sold. The EE value for the multipurpose trees was low. This is strange since one would expect a considerable pressure on trees for fuel wood considering the fact that this constitutes a significant energy source for the majority of the rural households in Africa (Kituyi et al., 2001). However it must be noted that the multipurpose trees were dominated by edge trees which, apart from acting as wind breaks, are only used occasionally for house construction or for fuel wood during festivities and funerals.

Summary statistics and agroecosystem performance indicators
The statistics calculated by Ecopath and system performance indices computed within and outside the model are presented in Table 5.3. The total system throughput

(sum of all imports, consumption, harvests and exports and returns to detritus) for the diversified integrated system was estimated as 162 kg N ha^{-1}yr^{-1}. The system harvest index decreased slightly upon introduction of Fingerponds probably due to increased system throughput whilst the net yield remained constant since all production from the new, integrated production system (Fingerponds) was consumed by the household. The benefit of including Fingerponds into the integrated farming system is seen in the decrease in the import fraction to the household, in which it plays a core role in the overall system resource flows. This is due to addition of fish and vegetable flows in the household diet composition. Increasing the area of the wetland used for Fingerponds increases the overall agro-ecosystem productivity. The mean trophic level of the system yield is 2.11, indicating dominance of biomass by level 1 consumers (in this case physiological primary producers). Generally, the B/E value is low depicting a characteristic of an immature ecosystem (Odum, 1969). The agroecosystem functional diversity increases with increased integration of wetland farming activities into terrestrial production systems, and also Fingerponds into the entire farming system and confirms the observation by Dalsgaard and Oficial (1997).

A scenario with the exclusion of the households within the model boundary indicated an increase in the net system yield by about 30%. This is the proportion consumed by the households. About 50 % of the total harvest is mainly in the form of papyrus mats on transit to the local markets. The overall agroecosystem harvest was 14 kg N ha^{-1}yr^{-1} of which about 3 kg N ha^{-1}yr^{-1} was the household consumption. There is however an additional external input is derived from the local markets through consumption of foodstuffs purchased from the proceeds of papyrus mat sales.

Figure 5.5 shows an Ecopath flow diagram for the farming system based on selected aggregate of 12 moderate rural households in Kusa. The boxes indicate the size of each compartment in terms of average standing biomass expressed as mg N m^{-2} yr^{-1}. Terrestrial and wetland cultivated area detritus constitute the main nitrogen pools supporting the first level consumers (mainly plants). At the higher consumer level, ruminants dominate the average standing biomass nitrogen pool. Apparently, the multipurpose tree biomass, particularly within the model boundary, was low probably due to dominance by relatively young trees. The harvest represents the flows out from the model boundary and is dominated by papyrus biomass mainly through sale of mats at the local markets. Other important flows are arrowroot and livestock sales. Imports into the system include both organic components (mostly household food) purchased from the market and natural inputs (nutrient inputs from natural processes such as dry and wet deposition, biological nitrogen fixation (BNF), and flood related nutrient inputs into the wetland). Pond soil also received nutrient import from livestock through pond manuring. Unlike the ordinary bioresource flow schemes, the Ecopath model flow diagram indicates all flow paths including feedback flows to detritus, hence provides a clearer picture of the system characteristics.

Table 5.3: Summary statistics and ecosystem attributes. Scenarios 0, 1,2,3, and 4 stand for a system without Fingerponds; with Fingerponds (one per household); with two Fingerponds per household; four Fingerponds per household; and without wetland activities, respectively. Scenario 5 excludes the households in scenario 1.

Attribute	Scenarios					
	0	1	2	3	4	5
Sum of all consumption (mg N m^{-2}yr^{-1})	7454	8223	8920	10123	5637	7925
Sum of all exports (mg N m^{-2}yr^{-1})	1817	1343	1356	1437	1015	1565
Sum of all flows into detritus (mg N m^{-2}yr^{-1})	5623	6637	7500	9385	3649	6568
Total system throughput (mg N m^{-2}yr^{-1})	14895	16204	17778	20947	10329	16057
Sum of all production (mg N m^{-2}yr^{-1})	5603	6184	6692	7568	3796	6181
Agroecosystem enterprise richness	13	15	15	15	7	15
Functional agroecosystem diversity	1.71	1.72	1.74	1.76	1.34	1.56
System harvest index	0.19	0.18	0.17	0.15	0.03	0.22
Total biomass (excl. detritus) (mg N m^{-2}yr^{-1})	4187	4221	4229	4247	3335	3928
Net yield (mg N m^{-2}yr^{-1})	1093	1093	1108	1131	123	1394
P/B ratio per yr	0.00	0.09	0.16	0.27	0.03	0.09
B/E ratio per yr	0.28	0.26	0.24	0.20	0.32	0.25
Import fraction to household (% of imported resource flows)	37.4	28	28	28	53.9	

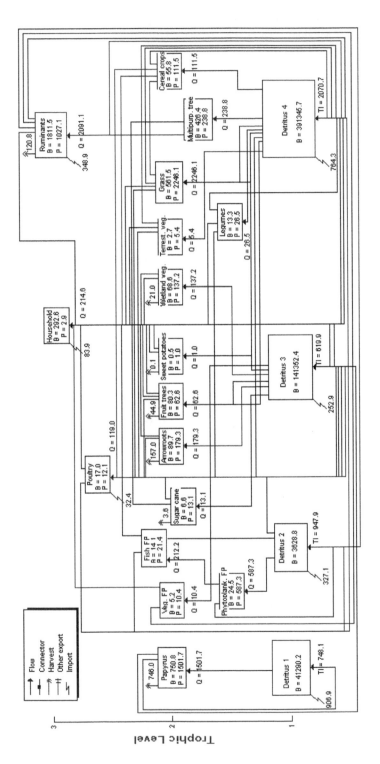

Figure 5.5: Ecopath model flow diagram of the Kusa agroecosystem

Table 5.4 shows the comparative computations of nutrient balances for the farming sub-systems at various levels. Most of the fluxes presented here were not measured *in situ* but are based on estimates from literature. At the Fingerponds level, the net nitrogen balance is positive. Anthropogenic input through pond manuring contributes to nutrient gains in the Fingerponds system.

Table 5.4: Comparison of estimated nitrogen balances (kg ha^{-1}yr^{-1}) in agroecosystems hierarchical levels across a hypothetical spatial scale

	Fingerponds	Maize (with legume intercrop)	Wetland cultivated area	Farming system level	Source of estimate
INPUT					
Livestock manure	222				This study
BNF (BGA)	47				Lin et al., 1988
BNF (symbiotic crops)		10	10	10	Brady, 1984
BNF (Asymbiotic fixation by Scattered trees and other plants)			2	2	Roy et al., 2003
Deposition (wet and dry)	4	4	4	4	Roy et al., 2003
Flood water	10		10	2	Roy et al., 2003
Incoming fish/seed	4	1		1	This study
Import from Market				1	This study
Sub total	**287**	**15**	**26**	**20**	
OUTPUT					
Gaseous losses	114	5	12	9	Hargreaves, 1998; Roy et al., 2003
Soil erosion		11		11	
Leaching	2*	2	1	2	Roy et al., 2003
Harvests	35	10	52	11	This study
Sub total	**151**	**28**	**65**	**33**	
Nutrient balance	**136**	**-13**	**-39**	**-13**	

*Estimates based on seepage loss estimates of pond water and average total nitrogen concentration from the seasonal water balance in Fingerponds (Chapter 3)

Discussion

Farming system diversification

Based on the approach used by Dixon et al., (2001) for classification of farming systems, the Kusa agroecosystem can be said to be a maize-mixed farming system incorporating root crops in the wetland. It is composed of diverse cropping systems with different sub-system combinations such as maize/sorghum/ legumes (beans and or groundnuts), fruit trees/banana/kitchen garden vegetable and wetland root crop/banana/sugar cane mosaic. Diversification of farming system enterprises both in terms of agricultural guilds and function not only leads to stability and security but also provides a balanced nutrition to the household. Integration increases bioresource flow and promotes not only the overall productivity but also the sustainable nutrient management as cycling pathways increase. Fingerponds increase the farming system diversity and intensification and may be seen as a livelihood improvement strategy. Traditionally, the diversity of these systems evolves with time as farmers attempt to buffer themselves from risks associated mainly with changes in weather patterns, diseases and pests, as well as the prevailing market forces. Over the past years the unreliable weather pattern has resulted in frequent crop failures (J. Odingo, pers. comm.) and farmers have been discouraged in putting additional investment in terrestrial production. For households in close proximity to the wetland, there has been a progressive shift of farming system enterprises from terrestrial to wetland-based production systems. Papyrus harvesting is a very important component of the entire system not only for nutrient flows but also for the household economic well-being. Wetland-cultivated crop harvests are used for both household consumption as well as supply to local markets. At the current production level and scale, Fingerponds are a subsistence enterprise and can only augment the household fish and vegetable supply.

The agroecosystem performance and nutrient balance

The nutrient throughput of the Kusa agroecosystem model is about 160 kg N ha^{-1}yr^{-1} and is generally low compared to over 350 kg N ha^{-1}yr^{-1} observed in a smallholder rice system in Philippines by Dalsgaard and Oficial (1997). Low throughput values are typical of agroecosystems with less open nutrient cycles. However, this feature may be influenced by other factors such as the nature of the farming enterprise (subsistence or commercial), and the definition of model boundary (e.g. the inclusion or exclusion of flows to the household(s) through direct consumption of the agroecosystem production). The inclusion of the households in the farming system model has advantages and disadvantages. On the one hand it improves our understanding of the human role in the agroecosystem functioning especially nutrient flows. Very often humans are treated as if they are separate entities from the agroecosystem upon which they exert considerable influence. Such an attempt also forms a basis for linking nutrient flows to household economics. This can then be used as inference for ecological/socio-economic dimensions; the corner stone of policy making. On the other hand, it complicates the interpretations of the agroecosystem nutrient flows. For instance, harvests from the agroecosystem that are consumed by households do not appear in overall system performance indicators.

The exclusion of households from the model (Table 3, Scenario 5) revealed the concealed agroecosystem harvests consumed within the model boundary. This allowed comparison with other farming system analysed in a similar approach using Ecopath. The system harvest was lower than the range of 30 to 65 kg N ha^{-1}yr^{-1} observed by Dalsgaard and Oficial (1997) in integrated and monoculture rice farms in the Philippines but was comparable with 13 kg N ha^{-1}yr^{-1}observed in upland integrated aquaculture in Quirino, Philippines (van Dam et al., 2002). The food products in the harvest amounted to 6.24 kg N ha^{-1}yr^{-1} in the Kusa farming system, lower but comparable with 7.59 kg N ha^{-1}yr^{-1} in the Philippines farming system. The main difference between the Kusa farming system and most Asian integrated agroecosystems is the low intensification and consequently low overall productivity per unit area in the former. A considerable proportion of land is under-utilized or lies idle most of the year.

Based on the current size of one Fingerpond system per household (about 400 m^2 wetland area), the integration of these systems into the existing crop and livestock-cum wetland biomass harvesting increases the total flow network by 8.2 % and subsequently decreases imports into the household by nearly 25%. The overall capacity of the integrated system flow network increases with increase in the size of Fingerpond per household. A comparative scenario analysis illustrates that excluding the wetland components from the model reduces the overall flow network capacity by 36% indicating the critical importance of the wetland in the overall nutrient flow of the farming system. This is also reflected in the difference between the system harvest index in less diversified integrated, purely terrestrial, crop-livestock system and more diversified integrated wetland-terrestrial production systems.

The nutritional importance of Fingerponds lies in their contribution to household protein supply mainly through fish consumption. Applying Scenario 1, where it is assumed that each household owns one Fingerpond, the per capita per season protein supply ranges from 215 g to 551g for a poor and a good yield, respectively. This is based on an average 7 persons per household and the assumption that all members are resident in the village. However, this is usually not the case and often one or two household members are not permanently resident in the village because of migration to nearby towns for employment. The actual per capita supply may therefore be higher than indicated. Based on FAO/WHO/UNU (1985) recommendations of daily protein requirement of an average 52 g for adults, this is 1-3 % of the daily supply. Improved management and consequently higher fish yields, as well as increased Fingerponds ownership per household (scenario 2 and 3), further improves the supply. In addition to the protein supply, vegetable harvests from the Fingerpond gardens provide vitamins and other essential elements to the households.

The productive capacity of the agroecosystem is an important feature that can be used to evaluate the system characteristics and can be quantified in both monetary and nutrient equivalents (Dalsgaard, 1998). In this paper the net yield, estimated as kg N ha^{-1}yr^{-1} is used as a measure of productive performance of the entire agroecosystem with Fingerponds. The wetland plays a crucial role in the overall productivity of the entire agroecosystem as the exclusion of the wetland components in the model reduces the net yield by nearly 90%. The B/E ratio calculated by Ecopath indicates the ability of the system to convert nutrients into biomass. According to the approach used by Odum (1969) in comparing ecosystem strategy at different successional stages, this ratio is expected to increase as ecosystems mature.

In this case study, the B/E ratio is generally low compared to a typical Asian integrated farming system such as that modelled by Dalsgaard and Oficial (1998). The Kusa agroecosystem is a traditional low input subsistence system. Contrary to our expectation, a scenario of increased wetland use for Fingerponds further decreases the B/E ratio. This observation could be explained by the fact that increased intensification leads to a faster increase of throughput compared to the total biomass increase rate. At the same time the P/B ratio appears to increase with an increase in the number of Fingerponds per household, again probably due to increased productivity rates from a very low production system compared to the total biomass. Unlike natural systems, where succession culminates in a stable ecosystem with optimum biomass, agroecosystems are influenced by humans through frequent harvesting and may behave somewhat differently. Nevertheless, Dalsgaard and Oficial (1997) showed that B/E increases with diversification and integration. The behaviour of the integrated system with Fingerponds as observed in this study is perhaps caused by the generally low overall productivity at the time of integration, particularly in the terrestrial ecosystem. The Finn's cycling index was nearly zero indicating negligible recycling. The introduction of Fingerponds promotes recycling through the use of livestock manure in pond fertilization.

Nutrient balance is an important aspect of farming system productivity and the overall sustainability. The nutrient balance for maize production, which is a dominant activity in the region, was negative. However, the estimated value is lower compared to observation of up to -88 kg N ha^{-1}yr^{-1} by Van den Bosch et al. (1998) in studies carried out in Kenya. In Kusa, most farmers hardly use any fertilizer or organic manure on maize fields because they believe that this causes the plants to wilt quickly during the intra-seasonal dry spells. The uncertainty in weather condition has discouraged the farmers from investing in terrestrial crop production leading to very low yields. For instance during this study the average maize yield was less than 0.5 ton per hectare per year. Terrestrial production is at subsistence level and hardly any products leave the household. In fact there is always a shortfall, particularly of cereals. Considering the seasonal wetland gardens and papyrus biomass harvesting, the net nutrient balance shows a more negative status compared with the sub-system level (Fingerponds and maize crop). This indicates that there is more harvest than input. On the other hand there are no visible indications of declining productivity (e.g. abandoned gardens) probably because there is still an adequate nutrient pool in the soil. The question is: how long can this be sustained amidst increasing dependence on wetland productivity for livelihoods? The overall farming system nutrient balance is slightly negative but low compared to -102 kg N ha^{-1} yr^{-1} obtained at farm level for Kisii district in Kenya by Van den Bosch et al. (1998). Again, this could be attributed to the fact that the production in the study area was at subsistence level with low input and consequently low production, and little surplus for the market except for the papyrus harvest. Although the nutrient balances presented here are based on relatively crude estimates, the computations confirm the characteristics of most agro-ecosystem productions in sub-Saharan Africa (De Jager et al., 1998; Roy et al., 2003).

In low external input production systems the gains from natural processes are important for the system productivity. The net nitrogen gains from natural processes estimated using secondary data indicated low contribution to the annual productivity. The question then is what sustains productivity in such low external input agroecosystem? This deserves further study. However, the probable explanation for the observed productivity may be attributed to the detrital organic nitrogen pool. In nature, the soil organic nitrogen pool is continuously released through mineralisation (Brady 1984). This pool is continually replenished by flows to detritus (unused production that is not removed from the system). A study of a fringing wetland by Mwanuzi et al. (2003) revealed that they are naturally net exporters of total nitrogen and organic matter. Further investigation is needed to conclusively estimate the role of natural gains in the overall system productivity.

Integration that enhances nutrient recycling may be beneficial to smallholder subsistence farmers as the majority cannot afford reliance on external inputs to sustain farm productivity. In order to improve the sustainability potential of the entire agroecosystem, there is a need for improvement of the agroecosystem productivity through more intensification particularly in the terrestrial ecosystem. This will enhance the productivity and reduce the pressure on the wetland. At the same time, a limited degree of intensification in the seasonal wetland farming systems may also help to slow down the increased conversion of the wetland emergent macrophyte zone into crop plots. Wolf et al. (2003) have shown that intensification of the present agricultural land is required to sustain the food demand and free more land for other purposes such a biomass production.

Conclusion

The results of this study highlight the importance of natural wetlands in nutrient flows within the farming systems adjacent to Lake Victoria. Although papyrus harvesting, mainly for mat-making, is not utilized directly by households it seems to play a big role in nutrient flows within the entire farming system. It shows a strong link between nutrient flows and household economy. This case study reveals that the overall farming system productivity is low. The wetland components, which constitute less than of the to tal area, support the bulk of the agroecosystem nutrient flows. For such a farming system to be sustainable, more effort has to be directed to increasing the overall productivity while focusing on integrated nutrient management. The positive nutrient balance at the Fingerponds level indicates potential nutrient sustainability. There is also an indication of nutrient accumulation in pond soils implying that nutrient-rich sediments can be used to enrich the vegetable gardens during the dry season. Alternatively, if the ponds do not dry up, manure inputs may be reduced in subsequent seasons. To promote release of nutrients from the sediments, frequent disturbance of the sediment will be required (Brummett and Noble, 1995). This can be combined with continuous partial harvesting by seining through the ponds. Fingerponds increases the agroecosystem enterprise richness and the overall farming system diversity. Diversification and linking farming systems components improves the overall productivity and creates an avenue to sustainability. The effect of the introduction of Fingerponds on the agroecosystem maturity indicators such as the biomass throughput ratio and the production biomass ratio is minimal. This may be partly because the overall farming system productivity is low and also the fact that Fingerponds are a small part of a larger farming system. There is need for intensification of the agroecosystem productivity in order to enhance sustainable food supply.

Incorporating households in the model can be a challenge in the interpretation of nutrient flows especially for subsistence farming as most flows are brought to a sudden halt. Nevertheless, such an approach provides a better understanding of the role of households in nutrient flows in agroecosystems and provides an opportunity to link nutrient flows and economic analysis. This study is the first application of Ecopath in farming systems analysis in East Africa. Biomass estimation of trees, shrubs and livestock was the main challenge in this study. Again, obtaining data from rural households, where there is no tradition of record keeping is a daunting task. There is a need for refinement of data collection methodology and for an improvement in household participation in the exercise. The sampling techniques developed should be versatile and at the same time provide more accuracy.

Acknowledgements

I wish to acknowledge the financial support from the European Union Fingerponds project Contract no.ICA4-CT-2001-10037. Additional funding for fieldwork was provided by the International Foundation for Science, Stockholm, Sweden and Swedish International Development Cooperation Agency Department for Natural Resources and the Environment (Sida NATUR), STOCKHOLM, Sweden, through a grant no W/3427-1.

References

Altieri, M.A., 2002. Agroecology: the science of natural resource management for the poor farmers in marginal environments. Agriculture, Ecosystems and Environment 93, 1-24.

Brady, N.C., 1984. The nature and properties of soils, 9th edition, Macmillan, New York, pp. 284-312.

Brummett, R.E., Noble, R., 1995. Aquaculture for African smallhoders. ICLARM Technical Report no. 46, 69 pp.

Christensen, V., Pauly, D., 1992. Ecopath II- a software for balancing steady-state ecosystem models and calculating network characteristics. Ecological. Modelling. 61, 169-185.

Christensen, V., Walters, C.J., Pauly, D., 2004. Ecopath with Ecosim: a users guide. Fisheries Research Centre Reports Vol. 12 (4) University of British Columbia, Vancouver, Canada, 154 pp.

Conway, R., 1987. The properties of agro-ecosystems. Agricultural Systems 24, 95-117.

Dalsgaard, J.P.T., 1995. Applying systems ecology to the analysis of integrated agriculture- aquaculture farms, NAGA, ICLARM, 18 (2), 15-19.

Dalsgaard, J.P.T., 1997. Tracking nutrient flows in a multi-enterprise farming system with a mass balance model (ECOPATH). In R.A. Morris (ed.) Managing soil fertility for intensive vegetable production systems in Asia. Proceedings of an international conference, 4-10 November 1997, Asian Vegetable Research and Development Centre, AVRDC Publication no. 97-469, pp. 325-343.

Dalsgaard, J.P.T., 1998. Monitoring and modeling agroecological sustainability indicators at farm level. In N.F.C Ranaweera, H.P.M Gunasema and Y.D.A Senanayake (eds.): Changing agricultural opportunities: The role of farming system approaches proceedings of the 14th International symposium of sustainable farming systems, Colombo, Sri Lanka, 11-16 November 1996, pp., 231-240.

Dalsgaard, J.P.T., Christensen, V., 1997. Flow Modelling with Ecopath: providing insights on the agroecological state of agroecosystems. In P.S. Teng, M.J. Kropff, H.F.M. ten Berge, J.B. Dent, F.B. Lansigan and H.H. van Laar (eds.) Applications of systems approaches at the farm and regional levels, Kluwer Academic Publishers, pp. 203-212.

Dalsgaard, J.P.T., Lightfoot, C., Christensen, V., 1995. Towards quantification of ecological sustainability in farming systems analysis. Ecological Engineering 4, 181-189.

Dalsgaard, J.P.T., Oficial, R.T., 1997. A quantitative approach for assessing the productive performance and ecological contributions of smallholder farms. Agricultural Systems 55(4), 503-533.

Dalsgaard, J.P.T., Oficial, R.T., 1998. Modelling and analyzing the agroecological performance of farms with ecopath. ICLARM Technical Report 53, 54 pp.

Dalsgaard, J.P.T., Prein, M. 1999. Integrated smallholder–aquaculture in Asia: Optimising trophic flows. In E.M.A. Smaling, O. Oenema and L.O. Fresco (eds.): Nutrient disequilibria in agroecosystems: concepts and case studies. CABI Publishing, Wallingford, UK, pp. 141-155.

De Jager, A., Kariuki, I., Matiri, F.M., Odendo, M., Wanyama, J.M., 1998. Monitoring nutrient flows and economic performance in African farming systems (NUTMON) IV Linking nutrient balances and economic performance in three districts in Kenya. Agriculture, Ecosystems and Environment 71, 81-92

Dixon, J., Gulliver, A., Gibbon, D., 2001. Farming systems and poverty: Improving farmers livelihoods in a changing world. Food and Agriculture Organization, Rome, and World Bank, Washington, DC.

FAO, 2001. Integrated agriculture-aquaculture: a primer. Fisheries Technical Paper No. 407. Food and Agriculture Organization of the United Nations, Rome. 149 pp.

FAO/WHO/UNU, 1985. Energy and protein requirements. Report of a joint FAO/WHO/UNU expert consultation. Technical Report Series 724. World Health Organisation, Geneva.

Fernando, C.H., Halwart, M., 2000. Possibilities for the integration of fish farming into irrigation systems. Fisheries Management and Ecology 7, 45-54

Garrow, J.S., James, W.P.T., Ralph, A., 1993. Human nutrition and diatetics. 9th edition, Churchill Livingstone, Edinburgh.

Gichuki, J., Dahdouh Guebas, F., Mugo, J., Rabuor, C.O., Triest, L., Derhairs, F., 2001 Species inventory and the local uses of the plants and fishes of the Lower Sondu Miriu wetland of Lake Victoria, Kenya. Hydrobiologia 458, 99-106.

Hargreaves, J.A., 1998. Nitrogen biogeochemistry in aquaculture ponds. Aquaculture 166, 181-122.

Haylor, G., Bhutta, M.S., 1997. The role of aquaculture in the sustainable development of irrigated farming systems in Punjab, Pakistan. Aquaculture Research 28, 691-705.

Jones, M.B., Humphries, S.W., 2002. Impacts of the C_4 sedge Cyperus papyrus, L. on carbon and water fluxes in an African wetland. Hydrobiologia 488, 107-113.

Kairu, J.K., 2001. Wetland use and impact on Lake Victoria, Kenya region. Lakes and Reservoirs: Research and Management 6, 117-125.

Kipkemboi, J., van Dam, A.A., Denny P., 2006. Biophysical suitability of smallholder integrated aquaculture-agriculture systems (Fingerponds) in East Africa's Lake Victoria freshwater wetlands. International Journal of Ecology and Environmental Sciences 32(1), 75-83.

Kituyi, E., Marufu, L., Huber, B., Wandiga, S.O., Jumba, I.O., Andreae, M.O., Helas, G., 2001. Biofuel consumption rates and patterns in Kenya. Biomass and Energy 20, 88-99.

Leung, W.-T.W., 1968. Food Composition Table for Use in Africa. Rome: FAO, and Washington: US Department of Health, Education, and Welfare, 306 pp.

Liang, W., 1998. Farming systems as an approach to agro-ecological engineering. Ecological Engineering 11, 27-35.

Lightfoot, C., Bimbao, M.A.P., Lopez, T.S., Villanueva, F.F.D., Orencia, E.L. A., Dalsgaard, J.P.T., Gayanilo, F.C., Prein, M., McArthur, H.J., 2000 Research tool for natural resource, management, monitoring and evaluation (RESTORE) Volume 1 Field guide, ICLARM, Penang.

Lightfoot, C., Roger, P.A., Caguan, A.G., 1993. Preliminary steady state models of a wetland field ecosystem with and without fish. In V. Christensen and Pauly, D (eds.): Trophic models of aquatic ecosystems, ICLARM Conference Proceedings pp. 56-64.

Lin, C.K., Tansakul, V, Apinhapath, C., 1988. Biological nitrogen fixation as a source of nitrogen input in fish ponds. In R.S.V Pullin, T. Bhukasan, L. Tonguthai and J.L Mac Lean (eds.), The second international symposium on Tilapia in aquaculture. ICLARM Conference Proceedings 15, Department of Fisheries, Bangkok, Thailand, and International Center for Living Aquatic Resources Management, Manilla, Phillippines pp. 1-6.

Luoga, H.P., 2005. Nutrient flows and ecological sustainability of Fingerponds in the wetlands of Lake Victoria, East Africa. MSc thesis, UNESCO-IHE, The Netherlands.

Mwanuzi, F., Aalderink, H., Mdamo, L., 2003. Simulation of pollution buffering capacity of wetlands fringing Lake Victoria. Environment International 29, 95-103.

Ntiamoa-Baidu, Y., 1998. Wildlife and food security in Africa, Food and Agriculture Organization of the United Nations (FAO) Conservation Guide 33, 110 pp.

Odum, E.P., 1969. The strategy of ecosystem development. Science 164, 262-270.

Odum, H.T., 1983. Systems ecology. John Willey and Sons, NY, 644 pp.

Pauly, D., Christensen, V., Walters, C., 2000. Ecopath, Ecosim and Ecospace as tools for evaluating ecosystem impact of fisheries ICES Journal of Marine Science 57, 697-706.

Polovina, J.J., 1984. Model of a coral reef ecosystems I. The ECOPATH model and its application to French Frigate Shoals. Coral Reefs 3(1), 1-11.

Prein, M., 2002. Integration of aquaculture into crop-animal systems in Asia. Agricultural Systems 71, 127-146.

Roy, R.N., Micra, R.V., Lesschen, J.P., Smaling, E.M., 2003. Assessment of soil nutrient balance, Fertilizer and plant nutrition bulletin 14, FAO, Rome, 110 pp.

Ruddle, K., Christensen V., 1993. An energy flow model of the mulberry dike-pond carp farming system of the Zhujiang Delta, Guangdong province, China. In V. Christensen and Pauly, D. (eds.): Trophic models of aquatic ecosystems, ICLARM Conference Proceedings pp. 48-55.

Salafsky, N., Wollenberg, E., 2000. Linking livelihoods and conservation: A conceptual framework and scale for assessing the integration of human needs and biodiversity. World Development 28(8), 1421-1438.

Schlaman, G., 2003. Quantifying ecological sustainability of IAA-systems in the southern part of the Mekong Delta in Vietnam, by using Ecopath. MSc thesis, Wageningen University, The Netherlands.

Tivy, J., 1990. Agricultural Ecology, Longman, London, 287 pp.

Van Dam, A.A., Chikafumbwa, F.J.K.T., Jamu, D.M., Costa-Pierce, B.A., 1993. Trophic interactions a in napier grass (Pennisetum purpureum)-fed aquaculture pond in Malawi. In V. Christensen and Pauly, D. (eds.): Trophic models of aquatic ecosystems, ICLARM Conference Proceedings pp. 65-68.

Van Dam, A.A., Lopez, T., Prein, M., 2002. Quantitative estimates of ecological sustainability in upland integrated agriculture-aquaculture systems in the Philippines. In: Proceedings of Tropentag 2002: International Research on Food Security, Natural Resource Management and Rural Development. University of Kessel-Witzenhausen, October 9-11, 2002.

Van den Bosch, N., Gitari, J.N., Ogaro, V.N,. Maobe, S. and Vlaming, J. 1998. Monitoring nutrient flows and economic performance of African farming systems (NUTMON), III Monitoring nutrient flows and balances in three districts in Kenya. Agriculture, Ecosystems and Environment 71, 63-80.

Wolf, J., Bindraban, P.S., Luijten, J.C., Vleeshouwers, L.M., 2003. Exploratory study on the land area required for global food supply and the potential global production of bioenergy. Agricultural Systems 76, 841-861.

Zamora, R.G., Baguio, S.S., 1984. Feed composition table for Philippines. Philippine Research Council for agriculture and Resources Research and Development, National Science and Technology Authority, Philippine National Feed Information Centre, University of the Philippines, Los Banos, Laguna, 264 pp.

Chapter

6

Integration of smallholder wetland aquaculture-agriculture systems (Fingerponds) into riparian farming systems at the shores of Lake Victoria, Kenya: socio-economics and livelihoods

Abstract

This chapter presents the results of experimental Fingerponds: an integrated flood recession aquaculture–agriculture production system at the Lake Victoria wetlands in Kenya. The overall aim of the study was to assess the potential of Fingerponds as a sustainable wetland farming system for improving food security at subsistence level and within the context of the existing livelihood activities. The contribution of this new activity to the rural household livelihoods was evaluated. The strength of this innovative technology lies in the enhancement of natural, human and social capital. Since the production level is intermediate, the benefits may not be high in the short-term perspective. Economic analysis showed that the gross margin and net income of Fingerponds is about 752 Euros and 197 Euros per hectare per year, respectively. This is about 11 % increase in the annual gross margin of an average rural household around Lake Victoria. The additional per capita fish supply is 3 kg per season or more from a 192 m^2 pond. The potential fish protein supply of 200 kg/ha is high compared to most existing terrestrial protein production systems. Fingerponds have the potential to contribute to household food security and livelihood. The results of the sensitivity analysis indicated that the biophysical variations, which may occur from one wetland to another, have implications on the functioning and consequently the economic performance of these systems. This reinforces the need for the integration of these systems into other household activities to buffer the household against potential risk.

Key words: Kenya, wetlands, integrated aquaculture production, socio-economic analysis, livelihoods, food security, Fingerponds

---

Publication based on this chapter:
Kipkemboi, J., A.A. van Dam, M.M. Ikiara, P. Denny. Integration of smallholder wetland aquaculture-agriculture systems (Fingerponds) into riparian farming systems at the shores of Lake Victoria, Kenya: socio-economics and livelihoods. The Geographical Journal (submitted).

Introduction

The majority of the rural populations around Lake Victoria depend directly on the immediate environment for their livelihoods. Over the recent years, degradation of both natural and agroecosystems and climatic uncertainty have led to increased vulnerability of rural populations to food insecurity, engulfing them in a vicious cycle of poverty and environmental degradation. According to Barbier (2000), the link between rural poverty and environment in Africa has its roots in land degradation. Gray and Moseley (2005) argue that wealth and pursuit of economic development are to blame for large-scale environmental degradation. With increasing populations in many developing countries and the resultant pressure on ecosystems, the probability of many rural communities falling into a "Malthusian" poverty trap is increasing. The challenge is to manage effectively the natural ecosystems to meet the growing human for livelihoods needs in a sustainable way. Rural communities living around natural wetlands in Africa consider themselves lucky as these ecosystems, by virtue of their high primary productivity and high soil moisture (even during dry seasons) can still provide goods and especially food (Silvius et al., 2000). However, over the years encroachment and loss of African wetlands is expected to have a profound impact on these riparian livelihoods. Poverty and environmental stresses continue to be the main causes of food insecurity in Africa (Misselhorn, 2005): a major challenge in the Millennium Development Goals. There is no single formula for breaking this cycle. However, integrated and sustainable technologies should be considered as tools in facing the challenge.

The Lake Victoria basin is endowed with diverse wetland resources. Fishery, livestock husbandry, rain-fed agriculture and wetland biomass harvesting form the main livelihood activities for the local communities. Fishery is not only the pivot of the economic activities in the Lake basin but also an important source of livelihood for the majority of the rural riparian communities. The lake basin supports directly or indirectly about 30 million inhabitants. In Kenya, the lake generates some four billion Kenya shillings (about four million Euros) in foreign exchange and over seven billion to the fishers (Gitonga and Achoki, 2004). Over the years, there has been a decline in the Lake's fishery due over-fishing and degradation of the environment (Kassenga, 1997; Okeyo-Owuor, 1999; Odada et al., 2004). The introduction of alien species has impacted on the Lake's fishery (Hall and Mills, 2000; Goudswaard et al., 2002; Aloo, 2003; Balirwa et al., 2003). For example, the introduced Nile perch has dramatically increased landings but has decimated the endemic fish population, which constituted the main shoreline fishery for the communities. Coupled with increased demand from the international fish trade, there is little fish left for local consumption as the local market fish prices are prohibitive (Abila, 2003). Terrestrial rain-fed agriculture around Lake Victoria has also become increasingly unreliable due to erratic rainfall. These factors have contributed to the poor economy and impoverishment of rural communities so that wetland seasonal cultivation has become more important to supplement declining terrestrial production.

Wetlands also provide local communities with goods which can be traded for cash. For instance, papyrus is a common product used for mat making and thatching (Denny, 1995; Mafabi and Taylor, 1993; Gichuki, et al., 2001), but papyrus swamps are shrinking as their margins are converted into seasonal crop production plots. If this trend continues at the current rate, the future livelihood of rural communities at the wetland edge is at risk. Smallholder integrated wetland aquaculture (Fingerponds) were trialed at the Kenyan Lake Victoria wetlands. Ponds were dug into the wetlands and used for fish production while the excavated soil was used to create raised bed gardens between the ponds. When several narrow channel-like ponds and the adjacent gardens are constructed close to one another at the lake/swamp or swamp/land interface, they appear like "fingers" penetrating into the emergent macrophyte zone. This is why they are called "Fingerponds". This is an innovative, semi-intensive technology aimed at enhancing wetland products based on its natural functions, fishery and agriculture. The novelty of Fingerponds is that the ponds are filled naturally with water and stocked with wetland wild fish during the flood season. The fish become trapped in the ponds during flood recession and manure and vegetable wastes from the adjacent village are used to improve the pond productivity. Locally demanded vegetables are grown on the raised beds. The advantage of this system is that it enhances diversity of products as well as synergy between different components of the farming system. For instance, pond water may be used to irrigate the gardens while the sludge from the pond bottom may be removed during the dry season and spread over the raised beds. The Fingerponds concept is similar to the Chinese dike-pond systems (Korn, 1996), Mexican 'hortillonages' (Micha et al., 1992), and agri-piscicultural systems developed in Rwanda (Barbier et al., 1985) where crop and fish production systems are integrated to produce synergy between the two systems.

This technology appears promising particularly with respect to enhancing food security for the resource-poor rural riparian communities living around seasonally flooded wetlands in Africa (Denny et al., 2006). There is a need for understanding the economic and livelihood potential of such systems in the context of the existing household activities. This will then form the basis through which such technology can be recognized as an option for wise-use of wetlands and hence be incorporated into the wetland policy framework. The objectives of this study were (1) to characterize general rural riparian households at the Fingerponds experimental sites, using Kusa in Kenya as a case study, in terms of socio-economic status; (2) to analyze the differences between households and to assess the contribution of Fingerponds to households livelihoods; and (3) to evaluate the economic and food security potential of this systems vis à vis other existing farming system activities.

Methods

The study sites

In Kenya two villages around Lake Victoria; Kusa and Nyangera were selected. The Nyangera Fingerpond site is located on the northern shores of the lake at the littoral wetlands sandwiched between Kadimu and Usenge Bay at S 0° 3 ' 55.9", E 34° 4 ' 52.2" while Kusa is adjacent to the Nyando wetland on the eastern shores bordering Nyakach bay on the Winam gulf at S 0° 18' 1.2 ", E 34° 53' 21.3". The Nyangera site is about 500 m from the lake's shoreline while Kusa is located about 4 km from the shore at the outer margin of the floodplain wetland of the Nyando river.

In each site, the local communities participated in the Fingerponds construction and co-management. The Nyangera site was co-managed by the Nyangera primary school administration while the activities in the Kusa site were conducted by the K'omolo women's group comprising of 12 households. This paper focuses on the Kusa site, where a detailed monitoring of the households was carried out.

Data collection and analysis

The approach undertaken in this study employs different but complementary research methods. The methods used by Norman et al. (1995) and Eaton and Sarch (1997) were applied. The sustainable livelihood assessment approach was used to assess the contribution of Fingerponds to the local community living status (DFID, 1998, www.livelihoods.org; Broklesby and Fisher, 2003). Household monitoring and economic evaluation was based on adaptation from the Research Tool for Natural Resource Management, Monitoring and Evaluation (RESTORE) field guide approach (Lightfoot et al., 2000).

Socioeconomic surveys

Semi-structured interviews, direct measurements and observations were used to collect information from the households. The survey questionnaires were developed with the aim of capturing information at two levels: one focusing on the heterogeneity of households of the larger community and another targeting individual households involved in the Fingerponds pilot project. A household was considered as a "group of individuals who live on the same farm, work together on at least one parcel of land and recognize the authority of a single head of household in major decisions relating to the farm enterprise" (FAO, 1999).

Baseline survey   An initial baseline survey was carried out in July 2002. This was a random sampling survey based on semi-structured interviews carried out along three transects to a radius of 5 km from the Fingerponds research site. Questionnaires were pre-tested and adapted prior to the actual survey. A total of 79 households were interviewed, in each case the respondent was the head of the household. In many cases this was the husband except in case of *de facto* or *de jure* female head.

Background survey at selected village households   In May 2004, 12 households involved in the co-management of pilot Fingerponds were surveyed using a participatory research approach. Prior to the survey and monitoring, a meeting was held with the households to explain the needs and the use of the information solicited from them. This enabled us to obtain a better rapport with the households and eased our access to information during the subsequent monitoring period.

Monitoring households   From May 2004 to February 2005, 12 households were monitored monthly through semi-structured interviews. The monitoring period coincides with the Fingerponds season, which starts after the annual floods, normally in April and May and ends before the next flood period (Chapter 3). A questionnaire aimed at capturing time allocation on various livelihood activities, consumption, income and expenditure during the month were designed, pre-tested and adapted before administration. For the household consumption and expenditure, the study focused mainly on household food items. To improve accuracy, questionnaire focused on the last 7 days at the time of each interview. However a question as to whether there was any major activity outside the 7- day period was also posed to the respondents in order to capture any important event during the month.

Multiple linear regression was used to evaluate the main factors that influence harvesting of wetland products by the households using level of education, age and gender of the respondents, household size and access to the wetland as independent variables.

Evaluation of Fingerponds impact on household livelihoods

In July 2004, a livelihood assessment based on DFID guidelines was carried out to evaluate the contribution of Fingerponds to the household livelihoods (Noble, 2004). In this approach, livelihoods are viewed in the context of a pentagon of household assets: social, financial, human, natural and physical capital. Group meetings and individual interviews were conducted with members from the households involved in the Fingerponds pilot project. The assessment was made in a collaborative approach where the participants learned to do the evaluation themselves. This involved building people's skills so that they are able to appraise the relative importance of their activities in reducing household vulnerability. The respondents were provided with charts with icons depicting the various farming system activities and were requested to rank by strokes the importance of each enterprise in scale of one to five scale, where, 1= low significance 2 = below medium significance, 3 = medium significance, 4= above medium significance and 5 = very high significance (Figure 6.1). The relative contribution of each activity to household livelihood assets was also evaluated using a similar approach.

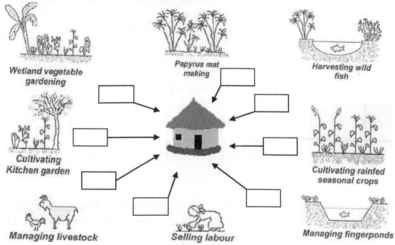

Figure 6.1: Livelihood activities ranking chart (adapted from Noble, 2004)

Economic analysis

The gross margin, net income and returns to labour were used as indicators of the economic performance of the various household enterprises. For convenience, the household enterprises were grouped into functional groups which form distinct management entities. The main farming system activities include: cereal and legume cultivation, livestock rearing and seasonal wetland agriculture. The dominant natural wetland biomass harvested is papyrus, which is mainly used for mat-making. For non-farm activities, the study focused on wages and sale of labour as the main sources of income. Table 6.1 shows a summary of the economic variables considered in the analysis. For each enterprise, the gross income (based on yields, market price and size/area of enterprise for production) and variable costs from inputs such as labour, seeds (Half the normal seed rates for intercrops, especially maize was assumed) and land preparation (normally hire of draught animal power for tillage) was computed. Fingerponds fish and vegetable prices were obtained from direct valuation of the harvested products by the local communities.

Daily local labour costs were estimated using opportunity cost of papyrus harvesting for mat-making. Mat-making involves cutting of papyrus culms, drying for 3-4 days at the site and knitting the mat. From our field data, observations and interviews, on average a household sells 6-9 mats every week at the local market. Using this information, the daily labour cost was estimated to an equivalent of 2 mats per person per day. Each mat costed 35 Kenya shillings (KES) at the local market. The daily labour cost is therefore estimated at 70 KES (about 0.7 euro at the time of this study). Harvesting is almost throughout the year except during floods. Based on our experience during pond construction, an average working day was estimated at 8 hours and may start as early as 6.00 a.m. The investment on the fixed assets were depreciated over their estimated useful period. For the Fingerponds a conservative lifespan of 10 years was assumed.

Table 6.1: Farming system economic variables and indicators

Attribute	Description
Gross income (GI)	Sum of cash income (products and by-products sold) and non-cash income (home consumption, household reserves, in kind payments, given away and farm use)
Fixed costs (FC)	Cost of infrastructure development + purchase of durable equipment
Variable costs (VC)	Inputs (seeds, fingerlings, manure etc) + Labour (paid and unpaid) + feeds and other charges e.g. Veterinary services)
Total costs (TC)	FC + VC
Gross margin (GM)	GI - VC
Net income (NI)	GI - TC
Returns to household labour	NI/ Total household labour in person days

Sensitivity analysis

Production systems that are solely dependent on natural events are often associated with risks and uncertainties. A sensitivity analysis was carried out to see the effect of water supply (mainly initial filling by flood), soil characteristics and fish yields on the economic performance of Fingerponds systems. Table 6.2 summarizes the scenarios and assumptions considered. The latter is usually determined by an array of biophysical conditions. In all cases it was assumed that the ponds were manured with cattle manure at a rate of 2500 kg/ha every two weeks and the functional period for fish culture was at least six months

Table 6.2: Summary of assumptions for different scenarios (the main variants considered are water supply, site soil characteristics and potential fish yields)

Scenario type	Flood/fish natural stocking	Soil suitability for vegetable production	Potential fish yield (kg/ha)	Fish yield category	Description (Observed /hypothetical)
1	Yes	Suitable	1068	Average	Observed average
2	Yes	Unsuitable	500	Poor	Poor site selection
3	No	Suitable	1068	Average	Good site but no flooding*
4	Yes	Suitable	1500	Good	Above average yield
5	Yes	Suitable	2000	Better	With improved skills and technology

In all scenarios, it was assumed that an average household own a Fingerpond (about 200 m^2 pond and a similar size of garden)

*In this scenario it was assumed that the natural fish stocking did not occur and fingerlings had to be purchased from local hatcheries

Results

Household characteristics and livelihoods

Household structure and community–wetland interrelationships
The households interviewed consisted of three clans: Kayano, Koyiegi and Kotiang, all from the Luo tribe, a sub-tribe within the nilotes living around Lake Victoria. Among the respondents, 49.5 % were women. Household sizes ranged from 2-20 with a mean of 6.68 persons each. Age distribution of the respondents ranged from 18-88 years while the education levels were 20.3% and 65.8 % for "no formal education" and "primary school level", and 12.7 % and 1.3 % for secondary and tertiary levels, respectively. The main occupations of the households are farming, mat making and petty trade. Table 6.3 summarizes the responses on the households' relationships with the natural wetland. From this the importance of these ecosystems to local communities' socio-economic status was inferred. According to the baseline survey, 92.3% of the respondents from 79 households interviewed indicated that they harvest products from wetlands. Only 29.9 % of the households obtained water for household consumption from the wetland. This was rather strange because there were no permanent rivers flowing through the village and from our observation there were also few wells. The introduction of rainwater harvesting techniques by a community project in Kusa seems to have had a profound impact on the community water supply. This may explain the low dependence on the wetland for domestic water consumption. The survey also revealed that some households have a long history of harvesting products from the wetland, starting as early as 1930. The majority of the respondents indicated that there were no taboos associated with wetland use, so technology associated with the wetlands is not likely to face social unacceptability.

Table 6.3: Responses to households-wetland interrelationship (n=79)

Attribute	% response	
	Yes	No
Free access to wetland	88.6	11.4
Crop production in wetland	62.0	38.0
Dependence on food produced from wetland	84.8	15.2
Dependence on natural wetland biomass harvesting	92.3	7.7
Membership to a social group involved in utilization of wetland	52.6	47.4
Worries about the future of wetlands	57.7	42.3
Knowledge on managed fish production in wetlands	30.4	69.6
Taboos associated with wetlands	7.6	92.4
Support concept of Fingerponds	97.5	

A multiple regression model with the harvest of wetland products by the households as dependent variable and log age and the education level of the respondent, log household size, dummy variable for gender and dummy variable for access to wetland as independent variable was significant (F-value = 3.097, P-value = 0.01). However, the explanatory power of the model was low with only 14.4 % of the variance explained. The regression coefficient for gender was significant and indicated higher dependence of female respondents on wetland resources compared to the male counterparts. In many rural homes, women are directly involved in household food provision and interact with the environment on a daily basis. The coefficient for the education level was also significant and positive implying that the dependence on the wetland increases with the increase in education level. One might expect the opposite as the more educated members of the society would be expected to access formal employment opportunities. However, it must be noted that the highest education attained by the majority of the respondents was primary and secondary level (78.5%) compared to 20% and 1.3 % for "no formal" and tertiary education. Owing to the current unemployment situation in Kenya, the majority of the respondents would still rely on the natural resources and the agroecosystem goods for livelihood.

Figure 6.2 shows the major items obtained from the wetland as revealed from the survey data. Direct use values range from products such as papyrus and fish to several food crops grown in the wetland to meet deficits in terrestrial production. Other use values include: livestock grazing, domestic water supply, recreation and wild game meat. Figure 6.3 shows the relative importance of various wetland commodities to household well-being. Cash, food and materials for house construction are among the important values of natural wetlands for rural communities around the shores of Lake Victoria. Papyrus harvesting for mat and craft making forms the backbone of the household economy. The food value is mainly associated with the cultivation of crops and the occasional post-flood wetland fishery. Wild game meat is now rare since most of the wild animals have been driven out by anthropogenic activities. However occasionally the village may feast from a kill of wetland wildlife such as hippopotamus (*Hippopotamus amphibious*). The materials for house construction are mainly emergent macrophyte biomass, particularly papyrus, *Phragmites* sp., *Typha* sp. and a number of small sedges which are commonly used as thatching material for rural huts. Papyrus and

Phragmites sp. have additional uses in house construction where they are used as rails to hold mud on the hut walls. Papyrus culms are used to make ropes to tie these rails firmly on the poles as a substitute for nails. The clay soils from wetland are mixed with cow dung and used in plastering the rural huts. Other products obtained from the wetland include water, medicine and biomass fuel.

The characteristics of the 12 households monitored intensively are shown in Table 6.4. The average household size was generally large compared to the national average of 5 persons per household but comparable to the entire area as observed in the community survey. The main household occupations were farming, mat-making, petty trade, informal sector employment and sale of labour by household members. Considering the potential average household labour force, it was found that about half the total household's population could provide labour needed for physically demanding activities ignoring the possibility of disability. However, only 30.8 % are always at home while the majority has temporarily migrated to nearby towns in search of employment but occasionally come home during the weekend or a few weeks in a year. This indicates that labour can be a limited resource. The literacy level is low, especially among the adults. This may be an impediment to the households' access to sources of livelihood outside the natural environment.

Table 6.4: Characteristics of 12 households involved in Fingerponds pilot study in Kayano village, Kusa, Kenya. Land size given as mean ± standard error. (n=12)

Characteristic	Attribute
Average household size (Number of people)	6.75
Average household composition by gender-Male (%)	45.68
Female (%)	54.32
Female headed households (*de jure*) (%)	41.67
Age range for household members (years)	1-55
Adults (above 18 years)	49.40
Average household land size (ha)	1.77± 0.44
Education No formal education (%)	25.93
Primary (%)	64.20
Secondary (%)	9.88
Tertiary (%)	0

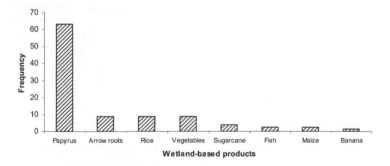

Figure 6.2: Major products obtained from the wetland by households

Households were grouped into three categories based on the typology used by Mutinda and Okotto (2001) for the same village. Three groups were designated as poor, middle and rich based on food supply, housing status, ability to provide basic necessities to their children and types of food consumed. The average total annual non-cash and cash incomes from the different farming system enterprises reflect the grouping of the 12 households into three relative classes: three poor, nine middle and one rich. Table 6.5 shows the relative contribution of various household activities to the household types. The dependence on wetland for both cultivated and natural wetland biomass is important to poor and middle class households while the rich family obtains significant non-farm income to meet household demands.

The results for the monthly average household time allocation indicated that livestock took the largest share of 33%, followed by 23 % and 15 % for non-farm activities and wetland crop cultivation, respectively (Figure 6.4). The time spent on livestock is mainly herding, normally freely within the household land or in the wetland as the land is not fenced; a common phenomenon in most villages around Lake Victoria. The children contributed almost half of the household time allocated to livestock. Terrestrial crop cultivation, natural wetland biomass harvesting and Fingerponds consumed 10%, 14%, and 5%, respectively.

Table 6.5: Relative importance of various income sources used as determinants of household socio-economic status (amount in Euros per year, mean EUR/KES exchange rate = 99.55)

Source type	Enterprise	Relative percentage contribution		
		Poor	Middle	Rich
On farm	Terrestrial crops	103.02	121.75	89.73
	Livestock	11.49	109.37	67.34
	Wetland cultivated crops	121.43	376.98	59.79
Non-farm	Natural wetland biomass	137.98	175.98	0
	Other non-farm	91.85	126.57	771.47
Total		465.77	910.65	988.33
Percentage of income from wetland sources		55.7	60.7	6.05
Percentage income from wetland natural biomass		29.6	19.3	0
Percentage non-farm sources over total income		49.3	33.2	78.1

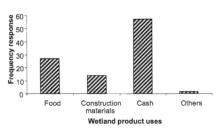

Figure 6.3: Main contribution of natural wetlands to households

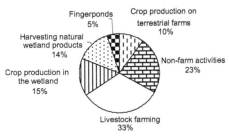

Figure 6.4: Time allocation on major household activities

Trends in natural wetland fishery and their implications

During the initial survey in 2002, 84.5% indicated that currently fish is not as easily available as before whilst 76.6% indicated that the commodity is no longer affordable. At the same time, 91.4% and 94.8 % of the respondents confirmed that in 1980s fish was easily available and affordable.

The results of fish consumption from the 12 households revealed that only 6% of the annual total household fish consumption was obtained from the nearby wetland while 94 % was from the market. The most popular fish among the households is tilapia (Figure 6.5). However, the increased demand from both the international and local market has driven prices high and tilapia is no longer affordable. The less desirable *Rastreneobola argentia* (popularly known as dagaa or *omena*) has taken over as the main fish consumed by households and currently constitutes about 60% of total household fish consumption.

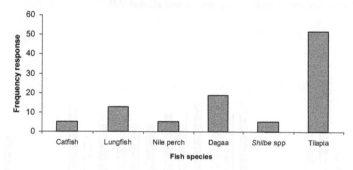

Figure 6.5: Fish preferences among the Kusa households community (n=79)

Household livelihood assets and contribution from Fingerponds
The results of the sustainable livelihood evaluation also confirmed the households' dependency on the natural environment to meet livelihood demands. The wetland is particularly important as half of the major livelihood activities rely on it. It must be noted that although some activities are ranked as important contributors to household well-being, this does not necessarily imply that they always provide adequately to the household requirements. For instance, rain-fed agriculture in the terrestrial part of the farms is a common practice in every household, however, due to erratic rainfall patterns, it has become very unreliable and crop failures are common. Similarly, wetland wild fish capture is short-lived and may vary between places and seasons depending on the size of the flood. On average, Fingerponds were comparable to most of the existing farming systems in terms of significance to household livelihood activities (Figure 6.6).

The common household activities have a variable contribution to various livelihood assets (Figure 6.7). Rain-fed agriculture and livestock, which is an old tradition of rural farming systems in Africa, still form the mainstay of people's household food security and well-being. Terrestrial crop production is viewed as a significant contributor to the financial resources not because of their significant sales but rather because of the savings on the cost of buying food, especially cereals, from the market.

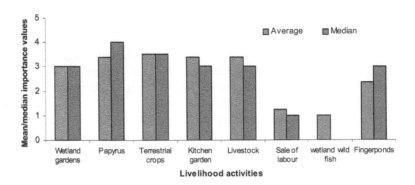

Figure 6.6: Ranking of importance of farming systems activities to household livelihood (1 = low significance 2 = below medium significance, 3 = medium significance, 4= above medium significance and 5 = very high significance). If an activity was not ranked in this scale then it was considered in significant and allocated 0.

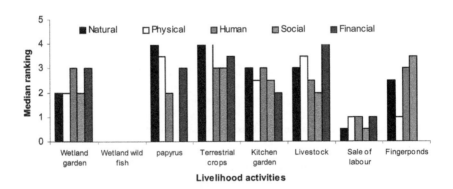

Figure 6.7: Importance ranking of the contribution of various household assets household livelihood (1 = low significance 2 = below medium significance, 3 = medium significance, 4 = above medium significance and 5 = very high significance). If an asset was not ranked in this scale then it was considered in significant and allocated 0.

Livestock and papyrus harvesting for mat making for sale ranked high in contribution to household financial capital. Normally the papyrus mats are sold in local markets which are held twice a week. Livestock is primarily considered a household investment and savings for large expenditures such as school fees, medical bills etc. Thus the objective of most households is to invest in livestock as a security. Accumulation of this investment may start from small animals such as poultry where combined sales can be transformed to small ruminants such as goats and sheep and then to cattle. The conversion of household animals to cash follows the reverse pathway depending on the scale of the need.

Reviewing Fingerponds against the so-called 'pentagon" assets of livelihood, it was found that the households viewed Fingerponds as an important contribution to their livelihood assets in the form of knowledge acquired; not only by the participants but also by interested members of the community who participated in various Fingerponds site activities. Working together among women involved in the project not only fostered inter-household relationships but also strengthened the social network, which is falling apart in many societies. In terms of the contribution to physical assets, the ponds were considered as an infrastructure development for harvesting of floodwater for fish production. The pond water can be used for irrigation of the vegetable gardens and also for domestic uses. The contribution to financial capacity was insignificant, probably due to the fact that these systems were mainly aimed at subsistence production and were operated with low input. However there may be an indirect impact when a household saves what would otherwise have been spent on vegetables and fish obtained from the market. Apparently wetland wild fish capture does not benefit the household livelihood probably due to the decline in catches. In general, it can be seen that rural households have very diverse livelihoods systems.

Economic analysis and contribution to household food security

The cost of adopting Fingerponds

Table 6.6 shows a summary of the cost of adoption for a Fingerpond at individual household level based on field experience in experimental systems in Kenya in 2002/2003. The main investment is in the construction; excavation of the ponds and spreading the soil to form the raised-bed gardens. Although there may be a high variability from place to place, the estimates obtained from construction in Kusa indicated that about 130 man days is required for one Fingerpond. For construction, the daily cost per man day was estimated to be equivalent to three mats per day (or an equivalent of KES 105) since the digging is more strenuous than ordinary farming activities. The estimated cost of Fingerpond construction is based on the assumption that the excavation of the ponds is carried out during dry season. Digging the ponds after the onset of heavy rains and when the wetland soil is saturated increases the overall cost by as high as two to three times the present estimate. This is because of additional costs like pumping out groundwater intrusion and extra man days for digging as the clayey wetland soils become heavy and difficult to work when wet. The cost of land has been ignored: the wetlands are state property and virtually free for use by the local rural communities. It is expected that the wetland policy will allow local communities to utilize the wetlands for livelihoods provided that their activities do not cause functional or structural changes. For example, the anthropogenic activities should impact minimally on the wetland hydrology.

The seasonal management requirements for Fingerponds can be divided into pond and garden management activities. On average, the pond management takes the largest share of about 47% of the total labour requirements compared to 31% and 22% for garden management and site clearing, respectively. Compared to the total construction costs, the annual labour costs for Fingerpond management are about 13%.

Table 6.6: Estimated cost of adoption for 1 Fingerpond (1 pond of 192 m^2 and a garden of the same area) in man days and costs (amounts in Euros per year, mean EUR/KES exchange rate = 99.55). Categories I and II are costs associated with the initial investment and the operating costs respectively.

Category	Activity	Description	Man days	Estimated cost
I	Construction	Cost of purchase of tools		24.12
		Initial site clearing	1.75	1.23
		Pond digging/soil levelling	127.9	134.90
	Other costs	Seine net (locally made)		40.18
Total				200.43
II	Pond management	Fish stock assessment and harvesting	2.44-6.12	3.01
		Manure collection and application	1.88-5.65	2.65
		Desludging	0.75-8.75	3.34
		Pond weed control	1.07-1.67	1.18
	Garden management	Land preparation	1.53-4.97	2.29
		Vegetable planting	0.31-0.91	0.58
		Vegetable watering	0.78-2.59	0.43
		Vegetable weeding	0.81-1.63	1.19
		Vegetable harvesting	0.39-1.63	1.40
	General site clearing	Clearing of emergent macrophytes around the Fingerponds	3.59-9.56	4.63
Total				20.70

Yields and economic performance of Fingerponds

The fish yields are variable between sites and seasons but on average 401.87 ± 26.03 and 1068.56 ± 99.35 kg ha^{-1} was attained after a 5-6 month growth period in manured ponds in Kusa and Nyangera, respectively (Kipkemboi et al., 2006). On average, fertilizing the ponds with cattle (*boma* or cattle enclosure) manure improves the fish yields to about one ton per hectare for the more suitable sites. Considering the fact that our technique for harvesting fish (seining through the pond) is partial since such systems cannot be drained, the potential yields and hence economic output could be slightly higher.

The vegetable yields of Kales (*Brassica oleracea),* a locally demanded vegetable, averaged 17 ton per hectare per season over two seasons in 2003 and 2004. The vegetable component of Fingerponds plays a very significant role in the economic performance of the system. The potential revenue from the vegetable constitutes about 70 % of the total Fingerponds system economic yield potential. This probably explains the relatively small difference observed in economic performance of manured and un-manured Fingerponds (Table 6.7). The vegetable gardens were not manured since the wetland soils were naturally fertile. Most of the vegetables were consumed by the households although under good management, surplus production can be attained and sold for cash.

Table 6.7: Gross margin of average productivity in Fingerponds systems with and without manuring (all units expressed Euros ha^{-1}, mean EUR/KES exchange rate = 99.55*)

Attribute	Manured	Un-manured
Gross income	1231.65	1079.34
Total variable costs	479.93	335.14
Gross margin	751.72	744.23
Total costs	1035.16	890.37
Net income	196.48	189.00
Returns to household labour (Euros/person day)	12.49	12.02

* Mean exchange rate at the time of this study (May 2004 - February 2005)

Comparative analysis of Fingerponds with other farming systems enterprises
Table 6.8 shows the economic performance indicators of various farming system enterprises. Papyrus harvesting appears to be the most attractive enterprise in terms of gross margin, net income and returns to household labour. Gross margin analysis of Fingerponds revealed that it is a viable enterprise and compares with arrowroot cultivation. However, a high fixed cost associated with the initial investment reduces the profitability to a net income of 72.8 % less than the gross margin. The gross margin of livestock is rather peculiar and cast doubts on the viability of this enterprise. Why do households still keep livestock then? To answer this question one needs to understand the role of livestock, particularly cattle, in a rural household set up in villages around the shores of Lake Victoria. The total gross margin and net income from all household enterprises was computed. Integration of Fingerponds into the existing farming system increased the gross margin and net income of the integrated system by 10.75% and 3.08 %, respectively.

Sensitivity analysis on Fingerponds economic performance
Table 6.9 shows a sensitivity analysis of the Fingerpond systems economic performance using a combination of observed and hypothetical levels of productivity. Poor site selection may lead to low yields due to biophysical limitations, especially soil mineral composition. There are patches of sodic soils at the wetland margin around Lake Victoria in Kenya (Mati and Mutunga 2003). Such soils may limit crop production in Fingerponds and may also limit pond productivity. In such cases the Fingerpond systems may yield negative returns in relation to the initial investment in pond construction. Additionally, even the sites with no limitations on soil conditions present some challenges. For instance, a site may not flood or the flood may be inadequate to stock the ponds. The farmer may be forced to obtain fingerlings from alternative sources such as local hatcheries and this increases the cost of production and reduces gross margin by 76.55 % compared to a situation when there is natural fish stocking.

Contribution to household food security
Fingerponds products consist of fish from the ponds and vegetables from the raised bed gardens. The monitoring study on household fish consumption revealed that the per capita annual fish consumption among the households is 5.03 kg with 4.7 kg sourced from the market and only 0.33 kg from the seasonal wetland fishery. Based

on pilot studies of joint ownership of four ponds (a total area of 768 m^2) by 12 households in Kusa, the average fish yields of about 1 ton per hectare per season supplied an extra 1.0 kg per capita fish per year. Assuming that individual households will own one Fingerpond of at least 200 m^2 and applying the average yields above and a household size of an average of 7 persons, the potential per capita fish supply is an additional 3.0 kg per capita per year. Under good management and effective final harvesting at the end of the season, a higher per capita supply can be achieved. The vegetables provided additional vitamins to the households.

Fingerpond's potential protein supply to households, particularly from fish production was compared with the other farming enterprises. Apart from arrowroots, whose biomass harvest per m^2 is higher than most of the cultivated crops, the potential protein supply from Fingerponds is about 200 kg per hectare and is higher than most of the other farming system enterprises (Figure 6.8). Cereals (predominantly maize and sorghum) constitute the main diet of many households in Kusa. However, cereal production is low due to multiple factors such as unreliable rain, low input and poor farming techniques.

The average food supply from existing terrestrial farming enterprises is low and hence households rely on food, particularly cereals, from other parts of the country. Contrary to our expectations, protein supply from ruminants appears to be negligible.

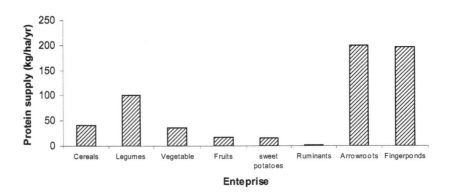

Figure 6.8: Comparison of on-farm protein supply from some household enterprises

Table 6.8: Comparison of the economic performance among the various farming system enterprises (all units expressed Euros ha^{-1}, mean EUR/KES exchange rate = 99.55*)

Enterprise	Gross income	Variable costs	Gross margin	Total costs	Net income	Returns to household labour per person day
Cereals	60.22	50.73	9.49	51.86	8.36	2.71
Legumes	173.64	112.80	60.85	113.74	59.91	5.33
Vegetable	255.90	113.16	142.75	117.62	138.27	10.72
Fruit trees	132.17	65.74	66.44	68.14	64.03	33.25
Sweet potatoes	165.75	122.94	42.81	129.52	36.23	4.16
Ruminants	66.31	137.47	-71.15	137.47	-71.15	-0.67
Arrow roots	1042.89	180.37	862.52	182.12	860.77	22.43
Sugar cane	164.14	55.95	108.20	57.14	107.00	20.84
Papyrus	2643.60	1215.04	1428.56	1219.99	1423.61	33.75
Fingerponds	1231.65	479.93	751.71	1035.16	196.48	12.49

Table 6.9: Fingerponds performance sensitivity analysis using different scenarios of fish production (all units expressed Euros ha^{-1}, mean EUR/KES exchange rate = 99.55*)

Scenario type[a]	Gross income	Total variable costs	Gross margin	Total costs	Net income	Net returns to household per person day
1	1231.65	479.93	751.72	1035.16	196.48	12.49
2	367.49	479.93	-112.43	1035.16	-667.67	-42.45
3	1231.65	1055.73	176.21	1610.66	-379.02	-24.09
4	1333.60	479.93	853.67	1035.16	298.44	18.97
5	1451.60	479.93	971.68	1035.16	416.45	26.47

[a]Description of the scenarios is provided in Table 2. * based on mean exchange during the mentoring period

Discussion

The link between wetlands and rural household economy

This study demonstrates the enhancement of wetland goods by building on existing uses for seasonal fishery and vegetable production. It reveals the degree of the dependence of rural households on wetland resources. There was over 90 % dependence on natural plant biomass (mainly papyrus culm harvesting) and 60 % on seasonal cultivated crops by the households. Schuyt (2005) indicated a similar high dependence of local riparian communities on natural wetlands in Yala swamp in Kenya. The indication of a significantly higher dependence on wetland resources by the female respondents shows that the majority of rural women, by virtue of their involvement in the day-to-day household food provision, tend to interact more with the immediate natural environment while men focus on cash income. The findings confirm the productivity potential of these ecosystems and their capability to support not only their endemic wildlife but also humankind and his livestock. It is under this premise that wetlands are regarded as economic strongholds for communities at the edge of these ecosystems (Crafter et al., 1992; Adams, 1993; Turner et al., 2000; Stuip et al., 2002). However, the factors that drive people's dependence on wetlands are not straightforward and are intricately intertwined with social, cultural and economic factors.

The wetland products harvested by rural households range from natural wetland biomass to seasonal agricultural crops. Over the last decade, the availability of fish has declined whilst the prices have increased leading to low affordability. This is one of the negative effects of international trade versus domestic supply (Abila, 2003). Small-scale fish production such as Fingerponds may therefore play an important role in the restoration of wetland fish supply to rural communities. Integration into the existing household activities creates a link between the wetland and the terrestrial livelihood production.

Technical and economic performance of Fingerponds

The results obtained in this study showed that with semi-intensive levels of management, Fingerponds can enhance wetland fish production. The natural fish stocking of these systems (Kipkemboi et al., 2006) saves about 55 % of what would otherwise be incurred in production costs. The cost of fry in a polyculture system in Sagana fish farm in Kenya constituted between 40-60 % of the total expenditure (Omondi et al., 2001). When compared with a situation where fingerlings have to be purchased, these systems are attractive and potentially more viable than the conventional systems. Manuring the ponds increases the fish yields: however, in terms of differences in gross margins of manured and un-manured ponds there was little difference. At the end of the culture period some of the fish were just table size (about 200 g) but the majority was smaller due to high fish densities in the ponds. The overall monetary value was therefore low. This lowers the cash potential of the Fingerponds fish component in the integrated systems to about 25 % of the total annual economic value.

On the other hand, focus on household food security indicates that Fingerponds have a high protein supply value (fish) while the vegetables may be more important for vitamins, fibre and carbohydrates. The vegetable biomass production of about 17 tonnes per hectare year is comparable to that reported in Nakivubo wetland, Uganda, of 15 tonnes (Emerton et al., 1999). According to FAOSTAT, the per capita freshwater fish supply in Kenya declined from 7 kg per capita in 1990s to a current low of less than 4 kg (Figure 6.9). Fingerponds provided an additional supply of 3 kg per capita at household level. This is equivalent to 18% of the world per capita supply average and 38% for Africa (FAO, 2004). The potential of Fingerponds to provide additional fish from wetlands is therefore a significant contribution towards the restoration of fish supply to households. This can also translate into savings as products from Fingerponds can substitute what otherwise would be purchased from the market.

The results showed that wetland-based enterprises such as papyrus, arrowroots and sugar cane had higher gross margins and net income than terrestrial systems. Although the gross income in Fingerponds per hectare is comparable to that of arrowroots, the cost of production associated with the initial investment in pond construction and purchase of fishing gear is relatively high. On average, Fingerponds increased the gross margin and net income of household enterprises by 10.75% and 3.08 %, respectively. The economic performance, particularly in the short term, can be higher if the initial cost of investment can be subsidized.

The poor performance of terrestrial enterprises such as cereal production is unusual and can be attributed mainly to poor rainfall. Indications of performance of livestock production activities is contrary to our expectation and does not agree with other studies in Kenya (c.f. De Jager et al., 1998). However, their studies were carried out in districts of high potential whilst ours was limited to basic benefits of traditional livestock breeds in the form of cash from sales and non-cash flows. Negative returns from livestock may appear misleading but in a short term analysis this tends to be the true picture compared to time invested in herding. In many rural communities, livestock is kept not only for consumable products but also for household finance, insurance and social status. The cattle breeds kept in most rural households around Lake Victoria are mainly the traditional East African zebu that are low milk yielders thus contributing little protein to the households. On the other hand these breeds are disease resistant and have low overall maintenance costs. Perhaps a comprehensive appraisal such as that given by Moll (2005) is needed to unearth these values.

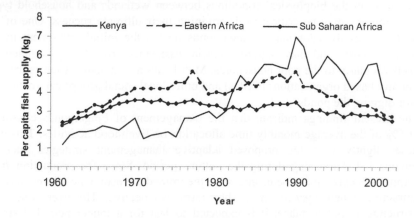

Figure 6.9: Per capita freshwater fish supply trends in Kenya, East Africa and Sub-Saharan Africa (Source: FAO, 2005)

Although our findings are based mainly on a detailed study carried out in one Fingerponds site, simple but indicative sensitivity analysis showed that under different biophysical conditions which are characteristic of different wetland sites, the economic performance of Fingerponds is likely to vary between sites and season. The hydrological pattern and its influence on the wetland flood regime also determine the performance of these systems (Chapter 3). Without natural floods at the Fingerponds sites there will be no fingerlings: this has a cost implication if the farmer has to source them from elsewhere.

Our approach to monitoring household activities, particularly in an African rural set-up, was an attempt to get an insight into the functionality of an integrated system. It gives a glance at the economic performance across various household enterprises particularly with respect to a new technology, "Fingerponds". Extrapolation of weekly information on household income, consumption, expenditure and time allocation to a whole month may not be of high accuracy. Moreover, it was realised that although most of the household information could be obtained from women, some, especially about non-farm income, is exclusively the preserve of men in male-headed households. There is a need to develop a simplified methodology for collecting socio-economic data that can be carried out by households themselves, with little training. Nevertheless our findings form a basis through which such an approach can be enhanced. A long-term evaluation framework is needed to capture the dynamics of adaptive management as these systems evolve over time.

The future of integrated aquaculture–agriculture systems in rural household livelihoods

Fingerponds should be viewed as diversification of wetland livelihood activities and not as a substitute. In this study, I have attempted to assess the potential of integration of wetland based aquaculture-agriculture systems into the existing households activities by looking at livelihood assets in totality. With many wetlands threatened by encroachment, particularly in sub-Saharan Africa, such a technology can provide the much-needed link between conservation and livelihood demands (Salafsky, 2000). The study demonstrates a high potential for Fingerponds although the variation in the biophysical conditions between wetlands and household types between regions may be critical in determining their ultimate success. One of the main challenges for adoption of Fingerponds lies in the initial investment in the construction costs and may require financial assistance from the government or NGOs particularly to the poor households. Mobilization of labour resources through community based organizations (CBOs) and other local social groups can reduce the cost for digging the ponds.

The present findings indicate that the management of Fingerponds consumes about 5% of the average monthly time allocation on livelihood activities. This may increase slightly with the proposed adaptive management strategies such as continuous fish harvesting during the culture period. Since the production from Fingerponds is extensive to semi-intensive, the immediate economic returns may not be attractive when viewed in a short-term perspective. However, once the infrastructure is put in place, it is expected to last for a longer period. Using a conservative estimate, a lifespan of ten years or more for a new Fingerpond can be achieved if well maintained. Thus it is reasonable to view the benefits in a longer perspective. Household livelihoods in rural communities in Africa are complex and require monetary valuation in order to fully appreciate and compare the performance across the diverse activities (Dovie et al., 2005). Robust methodologies for this type of analysis are needed. Unraveling this complexity requires understanding and adaptation to specific situations (Stroud, 2004).

Conclusions

The Fingerponds concept rekindles the wetland fishery, which has declined over the years and hence contributes to the development of natural resources. The technology has a potential role in improving the socio-economic and livelihoods of rural households living adjacent to natural seasonally-flooded wetlands. Considering the current levels of malnutrition of many rural children and women in Sub-Saharan Africa, Fingerponds have an important place in the fight against food poverty, particularly for communities living at the edge of seasonally-flooded wetlands. Crucially, they enhance food security through diversification of production and can be a potential lifeline in periods of stress or the dry season. Fish yields are important particularly for non-cash supply of food proteins to the household whilst the vegetables from the raised bed gardens, because of their high production, may more than provide for household requirements and the surplus can be sold for cash. The outcome of this technology therefore fits in the broader millennium development goal of fighting poverty.

Biophysical variations, which occur from one region to another, will be one of the main challenges to the performance of these systems. Due to uncertainties in rainfalls and floods, these systems should be operated as low input systems as much as possible. External input costs such as purchase of fingerlings and supplementary feeds should be avoided to keep them economically viable in a rural household setting. These challenges will require adaptive management and perhaps extension support from government or NGOs. To increase the viability of this technology, there is a need to reduce the cost and increase the productivity. The initial investment required for this technology, particularly pond digging may be a limitation in the poorest households. This requires institutional input particularly from the government, NGOs and other local community support groups, who will play a critical role in lifting this constraint. The construction can be done jointly while ownership can be individual or a group of few households. Again the layout and spatial location of the ponds in the wetland can be improved to enhance the chances of regular water supply and fish stocking by natural flood.

Acknowledgements

I wish to acknowledge the financial support from the European Union Fingerponds project Contract no. ICA4-CT-2001-10037. Additional funding for fieldwork was provided by the International Foundation for Science, Stockholm, Sweden and Swedish International Development Cooperation Agency Department for Natural Resources and the Environment (Sida NATUR), STOCKHOLM, Sweden, through grant no. W/3427-1. I would like to appreciate the advice by Dr Chiung-Ting Chang of UNESCO-IHE, The Netherlands on the economic analysis. I also wish to thank Dr. Reg Noble of International Support Group, Canada for the livelihood assessment and the local communities at the study sites for their participation and support.

References

Abila, R.O., 2003. Fish trade and food security; Are they reconcilable in Lake Victoria? Report of the expert consultation on international fish trade and food security, Casablanca Morocco 27-30 January 2003. FAO fisheries report No. 708, Rome.

Adams, W.M., 1993. Economy of the floodplain. In: G.E.Hollis, W.M. Adams and M. Aminu Kano (eds.) The Hadejia-Nguru wetlands: Environment, Economy and Sustainable Development in Sahelian floodplain Wetland. IUCN, Gland and Cambridge.

Aloo, P.A., 2003. Biological diversity of the Yala swamp Lakes, with special emphasis on fish species composition, in relation to changes in the Lake Victoria basin (Kenya); threats and conservation measures. Biodiversity and Conservation 12, 905-920.

Balirwa, J.S., Chapman, C.A., Chapman, L.J., Cowx, I.G., Geheb, K., Kaufman, L., Lowe-McConnell, R.H., Seehausen, O., Wanink, J.H., Welcomme, R.L., Witte, F., 2003. Biodiversity and fishery sustainability in the Lake Victoria basin; an unexpected marriage? Bioscience 53(8), 703-716.

Barbier, E.B., 2000. Economic linkages between rural poverty and land degradation: some evidence from Africa. Agriculture, Ecosystems and Environment 82, 355-370.

Barbier, P., Kalimanzira, C., Micha, J.-C., 1985. L'aménagement de zones marécageuses en écosystèmes agro-piscicoles. Le project Kirarambogo au Rwanda (1980-1985) ed. FUCID, Namur, Belgique 35 pp.

Brocklesby, M.A. Fisher, E., 2003. Community development in sustainable livelihoods approaches- an introduction. Community Development Journal 38(3), 185-198

Crafter, S.A., Njuguna, S.G., Howard, G.W., 1992. The sociological and economic values of Kenya's wetlands in Wetlands of Kenya; The proceedings of KWWG seminar on wetlands of Kenya, National Museums of Kenya, Nairobi, 3-5, July 1991, viii 99-107.

De Jager, A., Kariuki, I., Matiri, F.M., Odendo, M., Wanyama, J.M., 1998. Monitoring nutrient flows and economic performance in African farming systems (NUTMON) IV Linking nutrient balances and economic performance in three districts in Kenya. Agriculture, Ecosystems and Environment 71, 81-92.

Denny, P., Kipkemboi, J., Kaggwa, R., Lamtane, H., 2006. The potential of Fingerpond systems to increase food production from wetlands in Africa. International Journal of Ecology and Environmental Sciences, 32(1): 41-47.

Denny, P., 1995. Benefits and Priorities for wetland conservation. The case for national wetland conservation strategies. In: M. Cox, V. Straker & D Taylor (eds.) Wetland Archaeology and Nature conservation,. Proceedings of international conference on wetland archaeology and nature conservation, University of Bristol, 11-14 April, 1994. HMSO, UK.

DFID, 1998. Sustainable livelihoods guidance sheets. www.livelihoods.org

Dovie, D.B.K., Witkoswski, E.T.F. Shackleton, C.M,. 2005. Monetary valuation of livelihoods for understanding the composition and complexity of rural livelihoods. Agriculture and Human Values 22, 87-103.

Eaton, D., Sarch, M., 1997. Economic importance of wild resources in Hadejia-Nguru wetlands, Nigeria Creed working paper no 13 14 pp.

Emerton, L., Lyango, L., Luwum, P. Malinga, A., 1999. The present economic value of Nakivubo urban wetland, Uganda, National Wetlands Programme and IUCN.

FAO, 2005. FAOSTAT data, http://faostat.fao.org/ accessed February 2005.

FAO, 1999. Guideline for Agrarian systems diagnosis, FAO, Rome.

FAO, 2004. The state of world fisheries and aquaculture (SOFIA), FAO, Rome.

Gichuki, J., Dahdouh Guebas, F., Mugo, J., Rabuor, C.O., Triest, L. Derhairs, F., 2001. Species inventory and the local uses of the plants and fishes of the Lower Sondu Miriu wetland of Lake Victoria, Kenya. Hydrobiologia 458, 99-106.

Gitonga, N., Achoki, R., 2004. Fiscal reforms for Kenya fisheries in Cunningham, S.; Bostock, T. (eds./comps.) Papers presented at the workshop and exchange on fiscal reforms for fisheries to promote growth, poverty eradication and sustainable management, Rome 13-15, 2003, Fisheries Report no. 732, Suppl. FAO, Rome 19-29.

Goudswaard, P.C., Witte, F., Katunzi, E.F.B., 2002. The tilapiine fish stock of Lake Victoria before and after the Nile perch upsurge. Journal of Fish Biology 60, 838-856.

Gray, L.C., Moseley, W.G., 2005. A Geographical perspective on poverty-environment interactions. The Geographical Journal 171 (1), 9-23

Hall, S.R., Mills, E.I., 2000. Exotic species in large lakes of the world. Ecosystem Health and Management 3, 105-135.

Kassenga, G.R., 1997. A descriptive assessment of the wetlands of the Lake Victoria basin in Tanzania. Resource, Conservation and Recycling 20: 127-141

Kipkemboi, J., van Dam, A.A., Denny, P., 2006. Biophysical suitability of smallholder integrated aquaculture-agriculture systems (Fingerponds) in East Africa's Lake Victoria freshwater wetlands. International Journal of Ecology and Environmental Sciences 32(1), 75-83.

Korn, M., 1996. The dike-pond concept; Sustainable agriculture and nutrient recycling in China. Ambio 25(1), 6-12.

Lightfoot, C., Bimbao, M.A.P., Lopez, T.S., Villanueva, F.F.D., Orencia, E.L.A., Dalsgaard, J.P.T., Gayanilo, F.C., Prein, M., McArthur, H.J., 2000. Research tool for natural resource, management, monitoring and evaluation (RESTORE) Volume 1 Field guide, ICLARM, Penang.

Mafabi, P., Taylor, A.R.D., 1993. The national wetlands programme, Uganda. In: Davis, T.J (eds.) Towards wise use of wetlands, Wise use Project, Ramsar Convention Bureau, Gland.

Mati, B., Mutunga, K., 2003. Integrated soil fertility and management assessment of the Kusa profile, Lake Victoria basin. In: G Kimaru (ed.) Regional Land Management Unit Technical report 89 pp.

Micha, J.-C., Halen, H., Rosado Couoh, J.-L., 1992. Changing tropical marshlands into agro-pisciculture ecosystems. In: E. Maltby, P. Dugan and J.C. Lefeuvre (eds.) Conservation and development: the sustainable use of wetland resources IUCN Gland.

Misselhorn, A.A., 2005. What drives food insecurity in Southern Africa? A meta-analysis of household economy studies. Global Environmental Change 15, 33-43.

Moll, H.A.J., 2005. Cost benefits of livestock systems and the role of market and non market relationships. Agricultural Economics 32, 181-193.

Mutinda, T., Okotto, L., 2001. Report on outreach in Kusa pilot project-Harambee zone, 14-15th June 2001. Guiding systems consultants limited, Nairobi.

Noble, R., 2004. An evaluation of the impact of Fingerpond project (contract no. ICA4-CT-2001-10037) on target groups at two sites (Kayano village and Nyangera school) on the shores of Lake Victoria, Kenya. UNESCO-IHE, Delft, The Netherlands, unpublished report.

Norman, D.W., Worman, F.D., Siebert, J.D. Modiakgotla, E., 1995. The farming systems approach to development and appropriate technology generation, FAO Rome.

Odada, E.O., Olago, D.O., Kulindwa, K., Ntiba, M., Wandiga, S., 2004. Mitigation of environmental problems in Lake Victoria, East Africa: Causal chain and policy options analyses. Ambio 33 (1-2), 13-23.

Okeyo-Owuor, J.B.,. 1999. A review of biodiversity and socio-economics research to fisheries in Lake Victoria, IUCN.

Omondi, J.G., Gichuri, W.M., Veverica, K., 2001. A partial economic analysis for Nile Tilapia Oreochromis niloticus L. and sharptoothed catfish Clarias gariepinus (Burchell 1822) polyculture in Central Kenya. Aquaculture Research 32, 693-700.

Salafsky, N.,Wollenberg, E., 2000. Linking livelihoods and conservation: A conceptual framework and scale for assessing the integration of human needs and biodiversity. World Development 28(8), 1421-1438.

Schuyt, K.D., 2005. Economic consequences of wetland degradation for local populations in Africa. Ecological Economics 53, 177-190.

Silvius, M.J., Oneka, M., Verhagen, A., 2000. Wetlands: lifeline for people at the edge. Phys. Chem. Earth (B) 25 (7-8), 645-652.

Stroud, A., 2004. Understanding people, their livelihood system, demands and impact of innovations to advance development. Uganda Journal of Agricultural Sciences 9, 797-818.

Stuip, M.A.M., Baker, C.J., Oosterberg, W., 2002. The socioeconomics of wetlands, Wetlands International and RIZA, The Netherlands.

Turner, K.R., van den Bergh, J.C.J.M., Söderqvist, T., Barendregt, A., van der Straaten, J., Malby, E., van Ierland, E.C., 2000. The values of wetlands: landscape and institutional perspectives Ecological-economic analysis of wetlands: scientific integration for management and policy. Ecological Economics 35, 7-23.

Chapter

7

Smallholder integrated aquaculture (Fingerponds) in the wetlands of Lake Victoria, Kenya: assessing the environmental impacts with the aid of Bayesian networks

Abstract

The use of wetlands to meet livelihood demands may have some effects on the ecosystem integrity. This study evaluates the use of wetlands for Fingerponds (seasonal wetland fishponds integrated with vegetable production) using experimental sites at the shores of Lake Victoria in Kenya. The major concerns such as potential eutrophication of the wetland groundwater through leaching, changes in wetland species diversity and the potential negative effects on wetland hydrology are addressed. For eutrophication, field monitoring using physical parameter and nutrient measurement (soluble reactive phosphorus, total phosphorus, ammonium nitrogen, nitrate-nitrogen, and total nitrogen) was carried out. The study revealed that there was no evidence of nutrient leaching from the ponds into the immediate wetland groundwater. There was a significant difference between the nutrient concentrations in pond water and the ground water in the immediate wetland environment. The Bayesian network approach was used to evaluate the main environmental concerns. Based on an updated Bayesian model, the overall environmental impact of Fingerponds was rated as low to moderate. This study is based on short-term monitoring of the experimental Fingerponds. There is need for continued monitoring during the implementation phase.

Key words: Integrated wetland aquaculture, wetlands, environmental assessment, Bayesian networks, Lake Victoria-Kenya

Publication based on this chapter:

Kipkemboi, J., A.A. van Dam, P. Denny. Smallholder integrated aquaculture (Fingerponds) in the wetlands of Lake Victoria, Kenya: assessing the environmental impacts with the aid of Bayesian networks. African Journal of Aquatic Sciences (submitted).

Introduction

Human activities in natural wetlands have impacts on ecosystem health (O' Connell, 2003; Liu et al., 2004; Simonit et al., 2005). The health of a wetland can be defined as the ecological characteristics that ensure continued provision of goods and services for livelihoods but at the same time secure the integrity and sustainability of ecosystem function (nutrient and energy flows) and structure (species diversity and abundance). Changes in natural ecosystems may take a long time to be noticed, and efforts to reverse undesirable outcomes may be costly. Anthropogenic activities within and around wetlands continue to be the main threats to wetlands as they are resources of many interests.

In Kenya, the threats to wetlands fall into three categories: poverty-driven small-scale encroachment by the riparian communities eventually resulting in destruction of large portions of these ecosystems; economic development-driven reclamation mainly for large-scale agriculture, and the introduction of alien species that can have negative impacts on biodiversity. Lake Victoria and its adjacent wetlands are a testimony of the effects of human activities on natural ecosystems. The littoral and floodplain wetlands are under pressure due to extensive utilization for livelihood activities by the local communities. In the Yala swamp on the northern shores of Lake Victoria, the wetland is currently under threat by large-scale reclamation for crop production. In the lake, the introduced Nile perch *Lates niloticus* and two tilapiine (*Orechromis niloticus and O. variabilis*) species resulted in increased landings which benefited fish processing industries, but on the other hand led to the disappearance of endemic fish species and a concomitant decline in subsistence fishery (Halls and Mills, 2000). The invasion of the South American aquatic weed *Eichornia crassipes* not only threatened the transport functions of the lake but also caused significant changes in its ecology and fishery.

Natural wetlands play a significant role in the livelihoods of rural communities (Chapter 6). In the littoral wetlands of Lake Victoria seasonal cultivation is common along the wetland margins. Natural wetlands are also used for the traditional flood pool fishery and harvesting of natural plant biomass (Gichuki et al., 2001). As the flood pool capture fishery is short-lived, fish protein supply from the wetland is insufficient for most of the year. Human population and poverty in the region are both increasing; as a result, the pressure on natural wetlands is increasing. Developing sustainable aquaculture is one way of improving benefits from these ecosystems (Frankic and Hershner, 2003). The wetland fishery can be enhanced through careful manipulation of the natural productivity of the ecosystem.

Smallholder aquaculture-agriculture (Fingerponds) were trialed at the littoral and floodplain wetlands around Lake Victoria in Kenya. These are earthen ponds excavated in fringe wetlands at the swamp/land interface during the dry season. The soil removed is spread to create raised beds for vegetable production. They are called "Fingerponds" because from a bird's eye view, several of these narrow channel-like ponds appear like "fingers" penetrating into the emergent macrophyte zone. The ponds resemble natural flood pools traditionally used for wetland fish capture by local communities while the gardens are a continuation of the existing seasonal swamp margin vegetable patches. The ponds are stocked naturally by wild fish during annual flooding of the wetland, the fish culture and garden management starting after flood recession. Manure from livestock and vegetable wastes is applied to the ponds to stimulate the production of natural fish food while water from the ponds may be used for irrigation (Chapters 3 and 4).

Fingerponds have a potential for contribution to food security and poverty alleviation of the people living around seasonally-flooded wetlands in Africa (Kipkemboi et al., 2006). Like any farming activity they are bound to have some impact on the environment. The aquaculture component of Fingerponds particularly raises environmental concerns because of the known negative impacts of aquaculture on the environment (Tucker et al., 1996; Beveridge et al., 1997; Boyd and Massaut, 1999; Lin et al., 2001; O'Brien and Lee, 2003; Porrello et al., 2003). An immediate potential effect is cultural eutrophication arising from the manipulation of pond productivity through nutrient enrichment by addition of livestock manure. Another effect is associated with the potential changes in wetland vegetation as a result of cultivation of terrestrial crops and opening of the emergent macrophyte zone for colonization by non-native invasive species. By nature, many wetlands in sub-Saharan Africa are sources of human diseases such as malaria and schistosomiasis. There is a fear that the creation of ponds may create more habitats for the vectors hence aggravate the problem. Nevertheless, for communities who live around natural wetlands, human life remains interwoven with wetland functions and values, and the dependence on the natural environment for their day-to-day need is inevitable. The way forward may be to find a balance between livelihoods benefits and environmental costs. A qualitative risk evaluation can be used for the assessment and monitoring of the impacts of technology on the environment (Christine, 2003).

The assessment of the impacts of human activities, and concomitant trends in the natural ecosystems is challenging due to effects of uncertain events such as weather and interaction between components. This creates a matrix of complex interacting factors. Classical statistics do not provide an adequate framework for dealing with uncertainties and diverse data (ecological, economic and social variables). This is because variables are measured in different units and in some cases only qualitative information is appropriate. The interaction of variables is complex so changes in one may have a knock-on effect on the state of one or several others. Furthermore, there is no clear-cut agreement on what constitutes a sustainable ecosystem state. To overcome these difficulties an adaptive management approach, in which decisions are made based on continued accumulation of knowledge, is a feasible option.

Bayesian networks (Bns) are capable of accommodating such variability in data. They apply the probability rule developed by Thomas Bayes, an 18 century English clergyman, to obtain an outcome through which an inference can be made. Bns have been used mainly in medicine and artificial intelligence (Jensen, 1996). There has been a debate as to whether they are appropriate in the ecological domain (Dennis, 1996). However over recent years they have been used to address environmental issues (Varis, 1995; Ellison, 1996; Varis and Kuikka, 1997; Sadoddin et al., 2005) and have been applied to natural resource management planning (Bromley et al., 2005; Prato, 2005).

This chapter is an appraisal of the environmental concerns of Fingerponds as part of its evaluation and a basis for monitoring at the implementation phase. The overall aim of this study is to evaluate the potential impact of Fingerponds on the natural wetland environment and on the lives of the local people especially with the aim of understanding the potential of scaling up the technology. The objectives were: (1) to identify the main environmental and social concerns of introduction of Fingerponds into natural wetlands and (2) to quantify and assess the implications of these concerns using Bayesian networks

Materials and methods

Identification of the major environmental concerns and developing criteria for assessment

What are the negative environmental and social concerns of the introduction of Fingerponds into the natural wetland environment? In order to answer this question, the general threats of aquaculture which may apply to Fingerponds were reviewed (Boyd, 2003). The most probable threats considered in this study include:

1. Nutrient enrichment/effluents associated with the intensification of production, particularly in pond aquaculture;
2. Introduction of alien species to the wetlands. These may be plants associated with arable crop production activities in the Fingerpond gardens and fish cultured in the ponds;
3. Habitat degradation and loss;
4. Modification of wetland hydrology through drainage;
5. Proliferation of human water-borne and water-related diseases;
6. Conflicts in resource use.

Data collection and analysis

In this study the Bayesian network is used to evaluate the environmental impacts of a smallholder integrated wetland aquaculture-agriculture (Fingerponds). The data is based on experimental Fingerponds sites in Nyangera and Kusa around Lake Victoria in Kenya (Chapter 1). Data from monitoring was used to evaluate the state of the system.

Bayesian modelling approach

A Bayesian network is a graphical model (*Directed Acyclic Graph* or **DAG**) used to represent a complex system in which variables (nodes) are linked by means of probabilities (Jensen, 1996). A Bayesian network can accommodate diverse data in the form of probability values, and can deal explicitly with uncertainties. Just like classical statistics, a Bayesian network is a tool of analysis/thinking and often an aid in decision-making (Ellison, 1996). As opposed to dynamic models, Bayesian networks provide a static or snapshot representation for a given period of time.

Environmental systems are made up of numerous complex interacting factors such that capturing the states of all the factors is difficult. The Bayesian network uses probabilities of states at any given time to generate a probabilistic inference through which a decision can be made. Quantitative and qualitative information about the system was transformed into interactions and consequences. A Bayesian model is based on three elements:

i.   A set of nodes representing variables in the environmental system. These variables can be physical, social or economic. Nodes are assigned states, which can assume discrete or continuous values.

ii. Links representing causal relationship between the nodes. The links are arrows originating from the cause (parents) to the effect (child). The relationships between the variables are defined by conditional probabilities.

iii. Probabilities assigned to each node specifying the belief that a node will be in a particular state given the states of those nodes that directly affect it. These probabilistic beliefs can be used to generate Bayesian statistics, which can then form a basis for inference. The probabilities can be based on rating evaluations derived from empirical data, expert knowledge and historical knowledge of the wetland users and local communities.

Environmental (ecological and socio-economic) variables in the system were identified, defined, and a probability model was set up using the Netica^{TM} programme (http://www.norsys.com/). At the beginning, the variables were assigned states and prior probabilities. This is a peculiar characteristic of Bayesian analysis in that it allows the use of existing knowledge in combination with the data collected in the overall system evaluation. This implies that prior to data collection or experiment, a prior probability distribution of variables can be computed through empirical data available to the scientist or subjective information obtained through careful thinking about the situation by the researcher, guided by personal experience. The probability distribution is then updated as new evidence from empirical data, as well as subjective data, becomes available. Netica will find the beliefs of all other variables when the network is compiled. From this, a posterior probability is obtained via Bayes theorem (Figure 7.1). This outcome summarizes what is known from the prior information and is used to make probabilistic inferences on the likelihood states of the interesting variable(s).

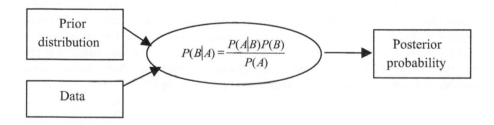

Figure 7.1: Data is processing using Bayes theorem

Bayesian statistics and inferences

The Bayesian approach to ecological analysis not only provides the model but also generates statistics which can be used to make inferences about the system in question. For instance, it provides a probability value for an environmental variable being investigated. However, Bayesian statistics are different from classical statistics and the two should not be confused when making inferences (Ellison, 1996).

The Bayesian approach in ecological analysis constitutes a radically different way of processing data. The fundamental difference between the Bayesian networks and classical statistics (frequentist) lies in the treatment of the parameters in question (Table 7.1). Classical statistics treat the system parameters as fixed values, meaning that variability can be accounted for within the confidence limits. However, this may not always be the case. On the other hand, Bayesian statistics treat system parameters as random distributions over the possible values while the data are fixed. Again the interpretation of probability under the two data analysis approaches differ. While the classical statistics treat probability as a result of a series of trials conducted under identical conditions, Bayesian statistics treat this as the observer degree of belief based on prior knowledge and data collected. Thus the conclusion of a P-value in classical statistic gives the probability of observing the results given a null hypothesis $p(x|H)$, while in Bayesian approach it implies the likelihood of an hypothesis, given data $p(H|x)$.

Table 7.1: A summary of inferences in the difference between classical and Bayesian statistics. (Adapted from Ellison, 1996)

Concept	Classical statistic interpretation	Bayesian statistic interpretation
Probability	This is the result of an infinite series of trials conducted under identical conditions	This is the observers degree of belief given data
Data	Random (representative) sample	Fixed (all there is)
Parameters	Fixed	Random
Conclusion	$p(x|H)$,	$p(H|x)$,

Identification of variables and definition of states
Wetlands are resources of many interests. Modelling all the uses and their implications is beyond the scope of this chapter. This chapter is confined to the use of wetlands for integrated aquaculture and the possible consequences. The variables were defined and assigned states (Table 7.2). The major threats were categorized into various likelihood degrees (Table 7.3). In this study the consequences of these threats focused mainly on the ecological characteristics of the wetland and the social aspects of the human population around it (Table 7.4 and 7.5). The social aspects were confined to human diseases and conflicts between the immediate wetland resource users, which may result from change in land use. Both impacts were combined to obtain an overall environmental impact rating (Table 7.6).

Table 7.2: Description of variables

No	Variable name	Definition	States
1	Fingerponds	Presence or absence of Fingerponds in natural wetlands.	Yes, no
2	Effluents	Nutrient enrichment or direct water pollution through effluents	Rare, unlikely, moderate, likely, almost certain
3	Species change	Threat to wetland structure due to introduction of species and un-intentional escape into the environment	
4	Hydrological modification	Modification of wetland hydrology particularly through drainage	Rare, unlikely, moderate, likely, almost certain
5	Human disease vector habitat	Creation of habitats for human disease vector organisms, e.g. mosquitoes (vectors for malaria), snails *Biomphalaria* sp. and *Bulinus* sp. (vectors for schistosomiasis)	Rare, unlikely, moderate, likely, almost certain
6	Change in land use	Change in land use through creation of aquaculture ponds versus the existing land uses	Rare, unlikely, moderate, likely, almost certain
7	Eutrophication	Risk of eutrophication particularly through loss of nutrients from the Fingerponds systems	Low, moderate, high
8	Biodiversity	Ecosystem degradation and negative effects on biodiversity	Low, high, moderate
9	Wetland hydrology	Potential negative effects on wetland hydrology	Low, high
10	Human disease	Whether Fingerponds may lead to proliferation of water borne and water related diseases among the adopters	Low, high
11	Resource use conflicts	Probability of Fingerponds leading to societal conflicts associated with resources use e.g. Fingerponds versus other existing activities such as biomass harvesting, livestock grazing	Low, high
12	Ecological impact	The impact of Fingerponds on the natural wetland ecological characteristics (structure and function) based on the nutrient status hydrology and species change.	Insignificant, minor, major
13	Social impact	The impact of Fingerponds on the socio-economic aspects (disease and resource use conflicts) to human communities living around them	Insignificant, minor, major
14	Environmental impact rating	Overall impact based on the ecological and social impacts	Low, moderate, high, very high

Table 7.3: Definition of states of variables (qualitative measures of likelihood of the impact)

Likelihood	Eutrophication	Species change	Modification of the wetland hydrology	Habitat for disease-causing organisms	Change in land-use
A. Rare	No discharge or leaching, but may occur	No risk of introductions of alien species	No interference of wetland hydrology	No creation of additional habitats for vectors	Intervention limited to harvesting of natural products at a sustainable level
B. Unlikely	Undetectable	Change limited to opportunistic invasion by seasonal short-lived terrestrial weeds	No significant modification of wetland hydrology	Some modification leading slight changes in existing habitats	Limited changes mainly within existing wetland uses
C. Moderate	Limited risk of loss of accumulated nutrients to the environment	Cultivation of new species or creation of new habitats, but with native species	Some modification of the hydrology with minor effects on water balance	Moderate enhancement of human disease vector habitats	Moderate change limited to intensive but small-scale intervention
D. Likely	Potential risk of water pollution through flushing out nutrient-rich water e.g. during flood	Introduction of new species from different environments	High degree of hydrological modification will ensue e.g. extraction of groundwater may be necessary	High proliferation of vectors due to presence of favourable habitat	Extensive detectable changes in land use
E. Almost certain	Discharge of nutrient-rich pond effluent into the environment through outflows/drainage	Introduction of alien species and genetically modified organisms (GMOs)	Hydrological modification cannot be avoided, e.g. channelization is inevitable	Alarming proliferation of vector organisms and evidence of threat to human health	Adverse changes in existing land-use in type and scale, often associated with a highly modified hydrological regime

Table 7.4: The magnitude of the consequences

Effect	Category	Description
Eutrophication	Low	No effluents or leaching of nutrient-rich pond water into the wetland groundwater or, if present, then at un-detectable levels
	Moderate	No detectable immediate effect but cumulative effect could occur
	High	Risk of nutrient enrichment of the immediate wetland environment evident
Wetland biodiversity	Low	No detectable change in wetland species diversity
	Moderate	Minimal changes of species diversity observed
	High	Evidence of potential or adverse changes in wetland species
Wetland hydrology	Low	No detectable change in wetland hydrology
	High	Construction and management of Fingerponds requires hydrological modification such as drainage channels
Human diseases	Low	Vectors of human disease may be present but abundance may not differ appreciable from that of the natural wetland
	High	Vector abundance present at alarming abundance
Resource-use conflict	Low	None or isolated cases of conflict between resource users
	High	High frequency of conflicts between the different user groups may occur

Table 7.5: Ecological and social impact rating

Rating	Ecological	Social
1. Insignificant	Minimal changes to the ecosystem function and structure. Effects may be undetectable and short-term	No significant change in natural habitat and land-use. Hence no impact on disease vector proliferation and resource-use conflicts
2. Minor	Minor changes in wetland ecosystem structure and function and minimal modification in hydrology.	Minor changes in land-use but within the existing functions. Activity does not lead to creation of new habitats for human disease vectors
3. Moderate	Medium effects with respect to changes in land-use, hydrological modification. If undesired changes detected, stopping them should enable the wetland to recover.	Measurable changes in land-use but does not lead to conflicts in resource-use. May increase habitats for human disease vectors but no creation of new niches.
4. Major	Major changes leading to modification of the hydrological regime, extensive change in land-use leading to waste production and consequently discharge to the surrounding environment and consequently habitat degradation and loss. In such a case recovery may not occur, or restoration will be expensive	Adverse change in land-use leads to conflicts among resource-users, creation of more habitats and new niches for human disease vectors and pathogens.

Table 7.6: Overall environmental consequence rating

Impact rating	Description
1. Low	Impacts on the environment are un-detectable and if present are usually short-term
2. Moderate	The intervention leads to medium environmental impacts but can be reversed if the activity is stopped. The impacts can be mitigated at low costs. For instance, should cumulative negative effects on wetland structure and function be detected, then abandoning the intervention should enable the wetland to recover naturally
3. High	Measurable effects on the environment. In this case, the activity leads to detectable environmental impacts and can escalate if scaled-up. The impacts can be mitigated but may require investment in rehabilitation.
4. Very high	Activity leads to significant environmental impacts ranging from ecosystem degradation to negative socio-economic effects. The environmental and socioeconomic costs outweigh benefits.

Based on this information, a preliminary Bayesian model was created by linking the related variables. The conditional probability table was constructed for each variable (Figure 7.2). To obtain an output, the preliminary Bayesian network was compiled and results of the model were used to infer the possible state of the variables of interest. New evidence required for updating or strengthening the model results was obtained from monitoring studies.

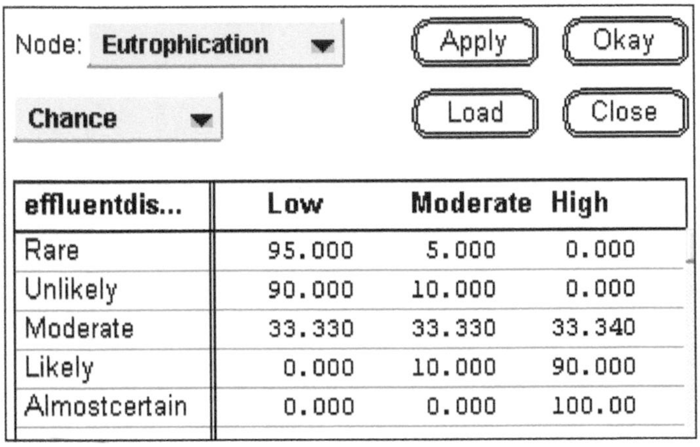

Node: Eutrophication ▼		Apply	Okay
Chance ▼		Load	Close

effluentdis...	Low	Moderate	High
Rare	95.000	5.000	0.000
Unlikely	90.000	10.000	0.000
Moderate	33.330	33.330	33.340
Likely	0.000	10.000	90.000
Almostcertain	0.000	0.000	100.00

Figure 7.2: Example of conditional probability table

Field data collection

Monitoring and quantitative assessment
The concerns relating to eutrophication and introduction of alien species were partly addressed through quantitative monitoring during the experimental period between 2003-2005. The potential contamination of wetland groundwater was assessed through monitoring of the physico-chemical parameters in the pond water and the immediate wetland groundwater. Some wetland plants are known to respond to nutrient enrichment through increased biomass and can be used as indicators of eutrophication (Kipkemboi et al., 2002). Vegetation biomass change was also used to monitor the potential eutrophication. For the risk of introduction of alien species in the wetland, the study concentrated on vegetation associated with the seasonal gardens since the aquaculture component relied on natural stocking from the wetland wild fish. During the experimental period, pond macroinvertebrates were monitored for the presence of disease vectors and other noxious organisms.

Pond – wetland groundwater water quality monitoring During the second season of the Fingerponds project (Chapter 3), three replicate wells of 1-2 metres deep were drilled around the ponds and in the surrounding (0.5, 1, 2, 5,6,7, 10, 20 and 50 m away from the ponds) wetland using a soil auger. The wells were left to stabilize for one day. During the second day, the wells were drained completely twice using a hand pump and allowed to recharge. After the water levels in the wells reached equilibrium with the groundwater, the water samples were collected for physico-chemical parameters and nutrient analysis. The assumption was that if leaching occurred from the ponds into the groundwater of the surrounding wetland, then this should be reflected in the similarity of the physico-chemical parameters.

Physical parameters (pH, electrical conductivity and total dissolved solids, and dissolved oxygen) were measured on site using a Jenway model 350 pH meter (Essex, UK), a Jenway model 470 conductivity/TDS meter (Essex, UK), and a portable dissolved oxygen meter model WTW-330i (Wissenchaftlich-Technische Werkstätten GmbH & Co., KG, Weiheim, Germany).

Samples for dissolved nutrients: soluble reactive phosphorus (SRP), NH_4-N and nitrate-nitrogen (NO_3-N) were filtered immediately through GF/C 47mm Whatman filters using Swinnex filter holders into acid-rinsed plastic sampling bottles. Un-filtered samples were also collected for total phosphorus and total nitrogen analyses. The samples were immediately stored under ice in cool boxes and transported to the laboratory for analysis following standard methods (APHA, 1992, 1995).

Vegetation studies In November 2004, a qualitative vegetation survey was carried out in the area impacted by Fingerponds and in a reference zone at a similar location on the wetland margin about 500 m away from Fingerponds. The purpose of the survey was to generate a species inventory to compare the potential impacted and non-impacted areas in terms of species composition. All species in transects of 100 m were identified and recorded. The un-identified specimens were collected and pressed for subsequent identification using references (Edwards and Bogan, 1951; Haines and Lye, 1983; Agnew, 1994).

Dominant emergent macrophyte biomass productivity was monitored for 8 months from February 2004. Permanent vegetation monitoring plots of 10 by 10 metres were demarcated in impact areas close to the ponds and selected control sites. At the beginning of the monitoring period all the above-ground plant biomass was harvested just above the soil level and allowed to regenerate. Three random quadrats of 0.5 by 0.5 m were used to harvest the aerial biomass within the marked plots. The harvested vegetation biomass was oven dried in the laboratory to constant weight at 60 °C. The dry weight was then determined and expressed as dry biomass in g/m^2.

Additional data During the study period, the pond macroinvertebrate community was monitored for the presence of human disease vectors. The potential conflicts emanating from the land-use for Fingerponds were also monitored. For instance in one study site, the livestock frequented the ponds in the afternoons and not only did they drink the water but also caused damage to the embankments. For the purpose of experiments, the Kusa site had to be fenced off to prevent livestock from accessing the ponds. In the Nyangera site (not frequented by livestock grazing except during the extreme dry season), the problem was encountered only a few times and stopped after complaints were raised with the individual livestock owners.

Statistical analysis of the physico-chemical parameters and vegetation data
T-tests were used to differentiate between the aerial biomass density of the dominant vegetation in the impacted and the control sites. ANOVA was used to compare differences in means of physico-chemical parameters in the ponds and the groundwater in the immediate environment. When the difference between the means was significant, the Tukey HSD test was applied to establish the homogeneous subsets.

Based on the information obtained from the monitoring studies, the beliefs were updated and a new network was compiled. The combined effect of the ecological

and social states was then used to infer the potential environmental impact of
Fingerponds.

Results

Initial Bayesian network

A preliminary Bayesian network was constructed (Figure 7.3). This formed the
baseline for the impact assessment. In the subsequent experimental period, the
evidence from monitoring studies on the issues of major concern was evaluated and
used as new evidence for the network updating.

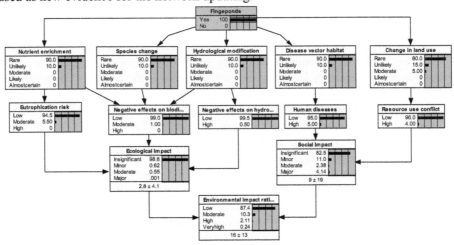

Figure 7.3: A preliminary Bayesian network

Evidence of environmental impacts

Eutrophication risk

Table 7.7 shows the physico-chemical parameters measured within the ponds and
the surrounding wetland environment in Nyangera, 2003. In Fingerponds, the risk of
direct effluent discharge is absent since there are no outflows from the ponds. The
analysis of the groundwater in the immediate wetland environment indicated that
there was no detectable risk of nutrient leaching from the ponds. There were
significant differences (P<0.05) in physico-chemical parameters between the pond
water and the wetland groundwater within 0.5 to 50 m, except for soluble reactive
phosphorus.

Table 7.7: Physico-chemical parameters variation in pond water and wetland groundwater at Nyangera Fingerponds. Values are means ± standard error (n=36 and 27 for pond water and the wetland groundwater respectively).

Parameter	Pond	Distance from Fingerponds (m)		
		0.5-2	5-7	10-50
pH (range)	8.10 – 9.85	6.23 – 6.77	6.23 – 6.69	3.61 – 6.85
Temperature °C	28.23 ± 0.23 [b]	25.9 ± 0.24 [a]	25.21 ± 0.27 [a]	24.60 ± 0.42 [a]
Dissolved oxygen (mg l^{-1})	13.95 ± 1.18 [b]	3.17 ± 0.24 [a]	2.99 ± 0.18 [a]	3.81 ± 0.33 [a]
EC mS cm^{-1}	10.68 ± 0.68 [b]	7.32 ± 0.47 [a]	6.42 ± 0.35 [a]	6.39 ± 0.54 [a]
Total dissolved solids (g l^{-1})	6.35 ± 0.40 [b]	4.43 ± 0.27 [a]	3.84 ± 0.21 [a]	3.81 ± 0.33 [a]
Soluble reactive phosphorus (mg l^{-1})	0.07 ± 0.01 [a]	0.03 ± 0.01 [a]	0.02 ± 0.00 [a]	0.03 ± 0.01 [a]
Total phosphorus (mg l^{-1})	0.11 ± 0.01 [ab]	0.06 ± 0.01 [a]	0.08 ± 0.01 [ab]	0.07 ± 0.01 [ab]
Ammonium nitrogen (mg l^{-1})	0.18 ± 0.02 [b]	0.04 ± 0.01 [a]	0.06 ± 0.01 [a]	0.04 ± 0.02 [a]
Nitrate nitrogen (mg l^{-1})	0.33 ± 0.30 [b]	0.04 ± 0.00 [a]	0.03 ± 0.00 [a]	0.08 ± 0.02 [a]
Total nitrogen (mg l^{-1})	1.52 ± 0.05 [b]	0.79 ± 0.05 [a]	0.78 ± 0.05 [a]	0.79 ± 0.07 [a]

Values in a row with the same superscript indicate homogenous subsets (Tukey HSD, P<0.05)

There was no difference between the nutrient levels in groundwater within the 0.5-50 m range, implying that the ponds had no direct effect on the wetland groundwater nutrient status.

Vegetation characteristics

Figure 7.4 shows the aerial biomass densities of the dominant aquatic macrophyte vegetation at the two study sites. In Nyangera, the vegetation exhibited zonation with papyrus dominating at the fringe, followed by a strip of *Typha domingensis*, and patchy stands of *Phragmites australis*. The vegetation at the periphery of the wetland adjacent to the Fingerponds consisted of a mixed stand of the three macrophytes. Besides the dominant vegetation, there was a variety of other plant species ranging from typical wetland vegetation to terrestrial invaders (Appendix 1 and 2). In Kusa, the vegetation was dominated by *Cynodon dactylon*, *Cyperus* spp. and *Cyperus papyrus*. *Cynodon dactylon* and *Cyperus* spp. dominated the immediate environment surrounding the ponds while papyrus was restricted to the inner part of the wetland. Appendices 3 and 4 show the species list at the impact and reference zone in Kusa. A summary of vegetation characteristics of the two sites is given in Table 7.8. The number of indicator species resulting from anthropogenic disturbance did not differ appreciably between the impact and reference sites. Nyangera site was generally richer in terms of the actual species recorded compared with the Kusa site.

Figure 7.5 shows the aerial biomass productivity of the dominant vegetation in the immediate environment around Fingerponds and control plots at the end of the 8-month monitoring period. In Nyangera the biomass densities were similar in the potential impact and control sites (Figure 7.5a). At the end of the monitoring period, the vegetation biomass at the potential impacted and the reference plots were not significantly different (T-test, P>0.05). However, in Kusa the biomass densities were only similar for the first three months of monitoring. The final biomasses at the end of the monitoring period were significantly different (T-test, P<0.05), the impact zone having a higher biomass than the reference (Fig. 7.5b).

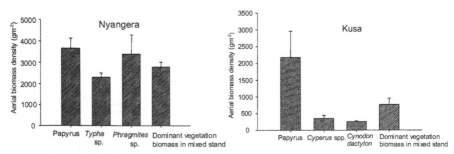

Figure 7.4: Aerial biomass density of the dominant vegetation from three random quadrats at the beginning of the study. Values are means ± standard error (n = 3).

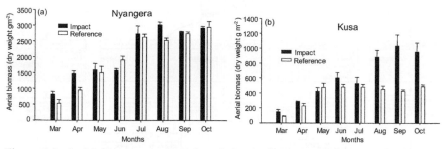

Figure 7.5: Aerial biomass productivity of the dominant macrophytes at the impact and reference plots in Nyangera and Kusa Fingerponds. Values are means ± standard error of dry weight biomass in three random quadrats.

Table 7.8: Vegetation characteristics in impacted and reference zones at the Fingerponds sites

Attribute	Nyangera		Kusa	
	Impact zone	Reference zone	Impact zone	Reference zone
Total number of plant species recorded	40	44	31	21
Number of species associated with anthropogenic disturbance	13	15	5	4

Detailed vegetation species list is provided in appendices 1-4

Updated Bayesian network and probabilistic inference
The potential overall impact of Fingerponds, based on ecological and social assessment is shown in Figure 7.6. This is derived by updating the Bn based on evidence gathered through field monitoring. The occurrence of most impacts was rated as rare or unlikely implying that they could occur at some time depending on Fingerpond management. Eutrophication of the adjacent wetland ecosystem through leaching of nutrients from the ponds did not seem to be a threat based on the monitoring studies and the fact that the ponds are un-drainable. Similarly the effects of Fingerponds on wetland hydrology were minimal based on the current design, which mimics the natural flood pools. However, there is a likelihood of some impact on the wetland vegetation structure through colonization of the seasonal wetland gardens with weeds of arable cultivation.

The occurrence of pond macro-invertebrates such as mosquito larvae and snails, which are intermediate hosts of human disease parasites, was low. The change in land-use from emergent macrophyte zone to a pond system can be said to be moderate (at subsistence level). The social risks associated with Fingerponds are low, except for potential resource-use conflicts. A combined effect of the potential conflict between livestock use and Fingerponds at the wetland margin may lead to a variable degree of social impact ranging from insignificant to minor. The overall impact can be said to be between low to moderate.

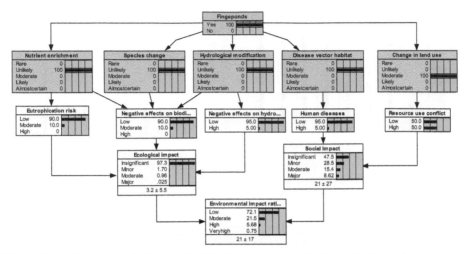

Figure 7.6: A Bayesian belief network of the Fingerponds implications for the environment
    after updating with evidence from monitoring

Discussion

Evaluating the environmental implication of integrated wetland aquaculture-agriculture

Eutrophication

The current design of Fingerponds is such that the water supply is dependent on
natural processes (Chapter 3). The design is such that there is no possibility of
draining the ponds, thus eliminating the risk of effluents. Fish harvesting is achieved
by either seining or the use of traditional fishing gear. Furthermore, the production is
aimed at a subsistence level and the fertilization of the ponds is aimed at maintaining
algal biomass productivity. This allows the application of two Best Management
Practices (BMPs) proposed for pond aquaculture by Boyd (2003): avoiding
excessive use of nutrients to stimulate pond productivity; and minimizing draining
of ponds during harvest.

There was no indication of nutrient loss into the surrounding wetland
environment through groundwater exchange. Papyrus, *Phragmites australis* and
Typha domingensis are all known to respond to nutrient enrichment by increased
biomass productivity and are often used in wastewater purification. These
macrophytes dominated the area around the Nyangera Fingerponds. The aerial
biomass did not differ between the impact and control zone implying that there was
no immediate eutrophication risk. Unlike Nyangera, the Kusa site was frequented by
livestock and the vegetation biomass in the control plots was affected by livestock
grazing during the dry season.

Another potential environmental concern is the accumulation of residues such as
antibiotics used in treatment of livestock by transfer to the ponds through manure
addition. However, according to the local veterinary official the use of antibiotics is
minimal since the dominant cattle breeds around Lake Victoria (mostly Zebu) are
disease resistant (Mr. E. Owuor, pers. comm.). The use of herbicides and pesticides

in the adjacent vegetable gardens was avoided. This was achieved by planting of disease-resistant crops and crops with less susceptibility to pest and disease attack.

Biodiversity

Around the shores of Lake Victoria in Kenya, undisturbed littoral wetlands margins are rare if present at all (J. Kipkemboi, personal observation). This is mainly due to the use of littoral wetland margins for various livelihood activities by the riparian communities. Weeds of arable farming were observed in the Fingerponds vegetable gardens. However, they do not pose a major threat as they cannot out compete the typical aquatic macrophytes especially if the hydrology of the wetland is not tampered with. The seasonal flooding of the wetland will keep the invasive terrestrial plants at bay. Under ordinary circumstances, a full re-colonization by aquatic macrophytes is expected if the farming is abandoned. Invasion of the newly created ponds by new macrophyte species was insignificant. In the Nyangera ponds in 2003, *Pistia stratiotes* was observed shortly after the ponds were disconnected from the floodwater but disappeared naturally during the fish culture season. The water lily *Nymphaea alba*, also not previously common in the emergent macrophyte zone but occasionally present in flood pools was common in the ponds and had to be controlled. In Kusa, the presence of water fern *Marsilea* sp. and *Ludwigia* sp. were observed, however they occurred only in small patches at the pond margins. All these species occur naturally within the littoral of Lake Victoria.

Fingerponds are stocked naturally with wild fish during the annual floods. In this way the introduction of species and subsequent impact on the natural genetic diversity through escape of alien species is minimized. The loss of biodiversity in the Lake Victoria fishery due to the introduction of alien species is a major concern (Aloo, 2003; Balirwa et al., 2003). The lessons learned from the Nile perch and the water hyacinth imply that the use of alien fish species or growing non-native macrophytes in Fingerponds should be prohibited.

Fingerponds and human health

Although natural wetlands provide numerous resources to humanity, they also serve as a habitat for human disease vectors. Alteration of the emergent macrophyte zone to create fish ponds may enhance habitats for disease vectors such as *Anopheles* spp. (malaria) and *Biomphalaria* spp. and *Bulinus* spp. (bilharzia). A high presence of culicid larvae in the ponds was observed shortly after flood recession but the population declined rapidly and remained low during the fish culture period. This observation could be attributed to predation by fish and other pond predators such as water bugs and beetles. The occurrence of schistosomiasis (bilharzia) vectors is common in the Lake Victoria region in Kenya (Karanja et al., 1997; Raahauge and Kristensen, 2000). The abundance of the above human disease vectors in the Fingerponds could be rated as rare to low. The distribution of these molluscs is mainly associated with the presence of aquatic weeds which provide them with substratum and shelter. Controlling the proliferation of aquatic weeds should be part of the pond management practice in Fingerponds and may help eliminate these vectors. Nevertheless, general health education and preventive measures should be an integral part of pond management.

Resource-use conflict

The creation of ponds at the land/swamp interface may create a potential conflict between livestock grazing and pond aquaculture. The ponds may turn into drinking points for livestock. In principle the two can co-exist if the number of livestock are few. Where there are many livestock, the damage to the embankments and water loss through drinking by livestock may be significant. One issue noted during this study which may become a source of conflict in the future is the ownership of and access to the wetland. Although the wetlands belong to the government, the boundary and access rights remain amorphous. This is one of the sources of occasional disputes. Further changes in land-use may further complicate the situation.

Application of Bayesian inference in impact evaluation and scaling up

Comparing the ecological and social risks of Fingerponds in the updated model, it appears that the ecological impact rating is low based on the experimental scale. However scaling up may increase the impact to minor/moderate. On the other hand, the impact on social aspects ranges from insignificant to minor. The overall rating of the environmental concerns of Fingerponds can be rated as low to moderate. Scaling up may increase the impact score and requires monitoring. The Bayesian network can be updated as new evidence emerges, hence it is a useful tool for adaptive management.

Scaling up of Fingerponds may present some new challenges. However some lessons learned from the experimental set up can be used as a basis for recommendation and monitoring in the implementation phase. A precautionary approach is required. Habitat destruction and loss may occur if large-scale production for trade is encouraged. To avoid haphazard conversion of wetland into ponds, a mechanism of prior authorization at the local level should be established. Wetland use may vary from place to place. There is a need for classification of wetland potential for Fingerponds (McCartney et al., 2005). This should be based on the potential environmental impacts and within the framework of a national wetland policy. Periodic review should be used to correct the negative effects, if detected.

Conclusion

Human activities in the wetlands, such as Fingerponds, have some environmental impacts. They may be positive, as in the case of increased food production, reduced dependence on wild fish stocks as well as diversified livelihood opportunities. In this case the utilization of wetlands for livelihoods is acceptable. The negative impacts are always of major concern. For instance, there are worries about the potential effect of scaling up and the perceived cumulative negative effects on wetland health. This may occur through the activities associated with the management of Fingerponds such as pond manuring and cultivation of terrestrial vegetables in the gardens. There was no evidence of immediate impact of eutrophication of the wetland through pond fertilization. The overall environmental impact of Fingerponds can be said to be low to moderate. Cumulative negative effects on the environment may not be manifested over a short period and constant monitoring is required. Stimulating the natural fish food through addition of just enough nutrients by manuring reduces the risk of eutrophication. Although

Fingerponds aim at enhancement of the existing wetland uses (flood pool fishery and seasonal agriculture), there may be some degree of change in land-use. Careful planning is required to achieve a balance between livelihood demand and maintenance of environmental integrity. This can be achieved by classifying the wetland potential for livelihood activities such as Fingerponds within the framework of a national wetland policy, which is currently in the draft stage in Kenya. Continuous monitoring and periodic review should be used to ensure sustainability.

Acknowledgement

I wish to acknowledge the financial support from the European Union Fingerponds project Contract no. ICA4-CT-2001-10037. Additional funding for fieldwork was provided by the International Foundation for Science, Stockholm, Sweden and Swedish International Development Cooperation Agency Department for Natural Resources and the Environment (Sida NATUR), STOCKHOLM, Sweden, through a grant no. W/3427-1. I wish to thank Patrick Moriarty of the International Water and Sanitation Centre (IRC), Delft, The Netherlands for guidance on Bayesian networks. I would also like to appreciate the assistance of Dr. S.T. Kariuki of Egerton University for assisting in plant species identification.

References

Agnew, A.D.Q., 1994. Upland Kenya wild flowers. A flora of ferns and herbaceous flowering plants of upland Kenya. East African Natural History Society, Nairobi.

Aloo, P. A., 2003. Biological diversity of the Yala swamp Lakes, with special emphasis on fish species composition, in relation to changes in the Lake Victoria basin (Kenya); threats and conservation measures. Biodiversity and Conservation 12, 905-920.

APHA, 1992. Standard methods for the examination of water and wastewater. 18th edition. American Public Health Association, Washington DC, United States of America.

APHA, 1995. Standard methods for the examination of water and wastewater. 19th edition. American Public Health Association, Washington DC, United States of America.

Balirwa, J. S., Chapman, C, A., Chapman, L. J., Cowx, I. G., Geheb, K., Kaufman, L., Lowe-McConnell, R. H., Seehausen, O., Wanink, J. H., Welcomme, R. L., Witte, F., 2003. Biodiversity and fishery sustainability in the Lake Victoria basin; an unexpected marriage? Bioscience 53(8), 703-716.

Beveridge, M.C.M., Phillips, M.J., Macintosh, D.J., 1997. Aquaculture and the environment: the supply of and demand for environmental goods and services by Asian aquaculture and the implications for sustainability. Aquaculture Research 28, 797-807.

Boyd, C.E., 2003. Guidelines for aquaculture effluent management at farm-level. Aquaculture 226, 101-112.

Boyd, C.E., Massaut, L., 1999. Risks associated with the use of chemicals in pond aquaculture. Aquacultural Engineering 20, 113-132.

Bromley, J., Jackson, N.A., Clymer, O.J., Giacomello, A.M., Jensen F.V., 2005. The use of Hugin® to develop Bayesian networks as an aid to integrated water resource planning. Environmental Modelling & Software 20, 231-242.

Christine, C., 2003. Qualitative risk assessment of the effects of shelfish farming on the environment in Tasmania, Australia. Ocean and Coastal Management 46, 47-58.

Dennis, B., 1996. Discussion: should ecologists become Bayesians? Ecological Applications 6(4), 1095-1103.

Edwards, D.C., Bogan, A.V., 1951. Important grassland plants of Kenya, Sir Isaac Pitman & Sons, Ltd, London, 124 pp.

Ellison, A.M., 1996. An introduction to Bayesian inference for ecological research and environmental decision making. Ecological Applications 6(4), 1036-1046.

Frankic, A., Hershner, C., 2003. Sustainable aquaculture: developing the promise of aquaculture. Aquaculture International 11, 517-530.

Gichuki, J., Dahdouh Guebas, F., Mugo, J., Rabuor, C. O., Triest, L., Derhairs, F., 2001 Species inventory and the local uses of the plants and fishes of the Lower Sondu Miriu wetland of Lake Victoria, Kenya. Hydrobiologia 458, 99-106.

Haines, R.W., Lye, K. A., 1983. The sedges and rushes of East Africa, A flora of the familes juncaceae and cyperaceae in East Africa-with particular reference to Uganda, East African Natural History, Nairobi, 404 pp.

Hall, S. R., Mills, E. I., 2000. Exotic species in large lakes of the world. Ecosystem Health and Management 3, 105-135.

Jensen, F.V., 1996. An introduction to Bayesian networks. University College London Press, London.

Karanja, D.M, Coley, D.G., Nahlen, B.L., Ouma, J.H., Secor, W.L., 1997. Studies on Schistosomiasis in western Kenya. I. Evidence for immune facilitated excretion of Schistosome eggs from *Schistosoma mansoni* and human imunodeficeieny co-infections. American Journal of Tropical Medicine and Hygiene 56 (5), 515-521.

Kipkemboi, J. van Dam, A.A., Denny, P., 2006. Towards sustainable community-wetland interaction: A pilot study on enhancing contribution to livelihoods through integrated aquaculture production systems (Fingerponds) at the Lake Victoria wetlands, Kenya. Paper presented at the wetlands, water and livelihoods workshop at St Lucia South Africa on January 30-February 2, 2006.

Kipkemboi, J., Kansiime. F., Denny, P., 2002. The response of *Cyperus papyrus* (L.) and *Miscanthidium violaceum* (K. Schum.) Robyns to eutrophication in natural wetlands of Lake Victoria, Uganda. African Journal of Aquatic Sciences 27, 11-20.

Lin, C.K., Shrestha, M.K., Yi, Y., Diana, J.S., 2001. Management to minimize the environmental impacts of pond effluent harvest draining techniques and effluent quality. Aquacultural Engineering 25, 125-135.

Liu, H., Zhang, S., Li. X, Lu, X., Yang, Q., 2004. Impacts on wetlands of large-scale land-use by agricultural development: the small Sanjiang plain, China. Ambio 33(6), 306-310.

McCartney, M.P, Musiyandima, M., Houghton-Carr, H.A., 2005. Working wetlands: Classifying wetland potential for agriculture. Research Report 90. Colombo, Sri Lanka: International Water Management Institute (IWMI).

O'Brien, P.J., Lee, C., 2003. Management of aquaculture effluents workshop discussion summary. Aquaculture 226, 227-242.

O'Connell, M.J., 2003. Detecting, measuring and reversing changes to wetlands. Wetlands Ecology and Management 11, 397-401.

Porrello, S., Tomassetti, P., Persia, E., Finoia, M.G., Mercetali, I., 2003. Reduction of aquaculture wastewater eutrophication by phytotreatment pond system II. Nitrogen and phosphorus content in macroalgae and sediments. Aquaculture 219, 531-544.

Prato, T., 2005. Bayesian adaptive management of ecosystems. Ecological Modelling, 147-156.

Raahauge, P., Kristensen, T.K., 2000. A comparison of *Bulinus* group species (Planorbidae, Gastropoda) by use of the internal transcribed trace 1 region combined by morphological and anatomical characters. Acta tropica 75, 85-94.

Saddodin, A., Letecher, R.A., Jakerman, A.J., Newham, L.T.J., 2005. A Bayesian decision network approach for assessing the ecological impacts of salinity management. Mathematics and Computers in Simulation 69(1-2), 162-176.

Simonit, S., Cattaneo, F., Perrings, C., 2005. Modelling hydrological externalities: the case of rice in Esteros del Iberà, Argentina. Ecological Modelling 186, 123-141.

Tucker, C.S., Kingsbury, S.K., Pote, J.W., Wax, C.L., 1996. Effects of management practices on discharge on nutrients and organic matter from channel catfish (*Italurus punctatus*) ponds. Aquaculture 147, 57-69.

Varis, O., 1995. Belief network for modelling and assessment of environmental change. Environmetrics 6, 436-444.

Varis, O., Kuikka, S., 1997. Joint use of multiple environmental assessment models by a Bayesian Meta model, the Baltic salmon case. Ecological Modelling 102(2-3), 341-351.

Appendices

Appendix 1: Plant species list for impacted zone in Nyangera

Species	Ecological status/ indicator of wetland disturbance/ specific notes
Abutilon mauritanium (Jacq.) Medic.	
Amaranthus hybridus L.	Common weeds in cultivated land
Aspilia pluriseta Schweint	Abundant in black cotton soils
Asystasia mysorensis (Roth)	
Bidens pilosa L.	Weeds associated with crop production
Brassica oleracea L.	Cultivated
Cleome monophylla L.	Weeds of cultivated and disturbed ground
Commelina banghalensis L.	Common and variable in bush land and disturbed habitats
Crotalaria polysperma Kotschy	
Cynodon dactylon Pers.	
Cyperus latifolious Poir.	
Cyperus rotundus L.	
Cyphostemma thomasii (Gilg and Brandnt.) Descoigns	
Dactylotenium aegyptium Beauv.	
Dichondra repens J.R. & G.Forst.	
Emex spinosus (L.) Campd	Local weed of cultivation
Euphorbia indica Lam.	
Felicia grantii (Oliv. & Hiem)	
Fimbristilis sp	
Hypoesthes forskahlii Vahl (R.Br.)	
Indigofera spinosa Fosrk.	
Ipomoea batatas (L.) Lam.	Originated from south America widely cultivated
Ipomoea cairica L.	
Ipomoea spathulata Hall.f.	
Leucaena sp.	Introduced fodder crop
Malva verticillata L.	Introduced, weed of waste places
Oxygonum stuhlmannii Dammer	Common in waste places and cultivation
Phragmites australis (Cav). Trin ex. Steud	
Schkurhia pinnata (Lam.) Tell	
Senna didymobotrya (Fresen.) Irwin & Barneby	
Senna occidentalis L.	
Sesbania sesban (L.) Merrill	
Solanum nigrum L.	Weeds of cultivation
Solanum sp.	Weeds of cultivation
Sonchus oleraceus L.	Weeds associated with arable land and gardens
Sphaelanathus cyathuloides O.Hoffm.	
Sporobolus sp.	
Triumfetta microphylla K. Schum.	
Typha domingensis Pers.	
Vicia villosa Roth.	Escape from upland cultivation
Other observed plants in the ponds	
Nymphaea alba L.	
Pistia stratiotes L. (only in 2002)	

Appendix 2: Plant species list for reference zone in Nyangera

Species	Ecological status/ indicator of wetland disturbance /specific notes
Abutilon mauritanium (Jacq.) Medic.	
Acacia pennata (L.) Willd	
Ageratum conzoides L.	
Archyranthes lanuginosa Schinz	
Aspilia pluriseta Schweint	
Brassica oleracea L.	Cultivated crop
Commelina banghalensis L.	Common and variable in bush land and disturbed habitats
Conyza bonariensis (L.) Cronq.	Weed of arable land cultivation
Crotalaria brevidens Benth.	Cultivated crop
Crotalaria italica	
Cynodon dactylon Pers.	
Cyperus papyrus L.	
Cyphostemma thomasii (Gilg and Brandnt.) Descoigns	
Cyprus rotundus L.	
Desmodium sp	
Dichondra repens J.R. & G.Forst.	
Digitaria sp.	
Fimbristilis sp.	
Hibiscus diversifolia Jacq.	
Hibiscus sp.	
Ipomoea batatas (L.) Lam.	Cultivated crop
Ipomoea cairica L	
Ipomoea spathulata Hall.f.	
Lantana camara L.	
Lycopersicon esculentum Mill.	Cultivated crop
Malva verticillata L.	
Manihot esculenta Crantz	Cultivated crop
Momordica calanthra Gilg.	
Phragmites australis (Cav). Trin ex. Steud	
Pluchea dioscoridis (L.) DC	
Polygonum pulchrum Blume	
Portulaca quadrifida L.	
Rhyncosia sp.	
Ricinus communis L.	Common in cultivated areas
Senna bicapsularis (L.) Roxb.	Common in disturbed dry bush land
Senna occidentalis L.	Common weed of cultivation
Sesbania sesban (L.) Merr.	
Sida rhombifolia L.	Common in disturbed places
Solanum incanum L.	Weeds of cultivation
Solanum nigrum L.	Weeds of cultivation
Triumfetta microphylla K. Schum.	
Typha domingensis Pers.	
Vicia vilosa Roth.	Escape from upland cultivation
Vigna unguiculata (L.) Walp.	Cultivated crop

Appendix 3: Plant species list for impacted zone in Kusa

Species	Ecological status/ indicator of wetland disturbance / specific notes
Abutilon longicuspe A.Rich.	
Abutilon mauritanium (Jacq.) Medic.	
Abutilon ramosum (Cav.) Guill & Perr.	
Acacia pennata (L.) Willd	Weed species in cultivated land
Amaranthus hybridus L.	
Barleria sp.	
Commelina banghalensis L.	
Crossandra tridentata Lindau	
Crotalaria brevidens Benth.	Cultivated crop
	Dominant pond edges and grassland
Cynodon dactylon Pers.	around the ponds
Cynodon plectostachium Pilger.	
Cyperus capillifolius A. Rich.	Invading pond bottom during dry season
Cyperus papyrus L.	
Cyperus rotundus L.	
Cyphostema thomasii (Gilg and Brandnt)	
Fimbristilis sp.	
Gomphocarpus physocarpus E.Mey.	
Hibiscus cannabinus L.	
Hibiscus sp.	
Ipomoea spathulata Hall.f.	
Leptochloa sp.	
Pluchea dioscoridis (L.) DC	
Polygonum pulchrum Blume	
Schoenoplactus sp.	
Sesbania quadrata Gillett	
Sesbania sesban (L.) Merr.	
Sida rhombifolia L.	Common in disturbed places
Solanum incanum L.	Weed of cultivated land
Spinacia oleracea L.	Cultivated crop
Sporobolus marginatus Hoechst	
Vossia cuspidata (Roxb.) Griff.	Invading in the Fingerpond area

Other plant species observed in the ponds
Marsilea sp. (2004)
Ludwigia sp. (2003)

Appendix 4
Plant species list for reference zone in Kusa

Species	Ecological status/ indicator of wetland disturbance / specific notes
Abutilon cannabinus L.	
	Common and variable in bush land and
Commelina banghalensis L.	disturbed habitats
Conyza bonariensis (L.) Cronq.	Weed of arable land cultivation
Cynodon dactylon Pers.	
	Seasonal wet grasslands, swamps, and
Cyperus rotundus L. subsp. B	waste places
Cyperus latifolius Poir.	
Cyperus papyrus (L.)	
Cyperus rotundus subsp. A	
Gomphocarpus physocarpus E.Mey.	
Ipomoea sp.	
Ipomoea spathulata Hall.f.	
Kyllinga sp.	
Phragmites australis (Cav). Trin ex. Steud	
Pluchea dioscoridis (L.) DC	
Polygonum pulchrum Blume	
Schoenoplectus sp.	
Scutelaria paucifolia Bak.	
Solanum nigrum L.	Weeds of cultivation
Trichomeria macrocarpa (Sond.) Hookf	
Vernonia lasiopus O.Hoffm.	Abundant in abandoned cultivation

Chapter

8

Synthesis and conclusions

Introduction

Wetlands play an important role in food production for rural households in Africa (Bugenyi, 2001). This study evaluated the potential of experimental aquaculture farming systems (Fingerponds) at the Lake Victoria wetlands in Kenya. These are flood recession integrated aquaculture systems. In order to understand how this innovative technology works, a functional (interaction between the ecosystem characteristics, structure and processes) and socio-economic approach was used. Analytical tools were employed to evaluate the different aspects of the technology within the farming systems. The biophysical suitability of the technology was assessed based on the existing wetland uses, soils and flood cycle. Pond water supply was examined using a dynamic simulation model. An Ecopath model helped evaluate the nutrient flows at the agroecosystem level. Cost benefit analyses and sustainable livelihoods' assessment were used to study benefits to households while the implications for the environment were assessed using a Bayesian network.

The results in Chapters 2 and 3 showed that the critical biophysical aspect, besides soil and nutrient inputs, was water supply. Pond hydrology was highly influenced by annual floods, the pond volume and the prevailing weather conditions after flood recession. The fish yields represented a considerable improvement over the capture fishery yields (Chapter 4). The nutrient flows analysis in the agroecosystem, using nitrogen as the currency, indicated that the flow capacity network (total system throughput) and hence the overall productivity, was low (Chapter 5). Fingerponds increase the richness and diversity of the farming system. The contribution to household livelihood assets and returns to labour were comparable with other household activities, and hence, integration into the existing activities can improve the overall benefits from the natural environment (Chapter 6). Unlike conventional aquaculture, Fingerponds take advantage of some of the existing natural processes and minimize the need for external inputs such as seed fish and feeds. They have minimal impacts on the environment, but there is need for continuous monitoring (Chapter 7). Careful planning of implementation of this technology will ensure the balance between livelihoods and conservation.

This chapter gives an overview of the potential of integration of Fingerponds into the existing farming systems around Lake Victoria. The first part addresses the functional aspects of the technology. In the second section, the potential benefits and recommendations for management are given. The last section of the chapter presents some policy issues, challenges, and conclusions.

The key determinants of Fingerponds functioning

Integration into the existing smallholder farming activities

Like in many rural areas of Kenya, smallholder farms around Lake Victoria comprise several components such as agriculture, livestock and poultry. The components can be integrated with wastes from one component serving as inputs into another. This leads to synergism. Fingerponds can be integrated into this system leading to more synergy (Figure 8.1). Around Lake Victoria the main farming activities are in the terrestrial ecosystem, but many farmers have to augment their production by harvesting natural biomass (mainly papyrus culms for mat making) and seasonal vegetables in the wetland. Such farming systems have evolved over time reflecting local indigenous knowledge developed through adaptive learning. In many Kenyan rural communities subsistence agriculture and animal husbandry are the main forces driving the rural economy. The main challenges of rain-fed agriculture around Lake Victoria in Kenya are the climate and the geology of the region. Most of the land around the lake is semi-arid and rocky. This forces local people to rely more on the wetlands when rains fail. The wetlands offer residual soil moisture upon which seasonal crops can be grown and are also a source of fish and biomass, e.g. papyrus culms for mat-making. However, the wetland capture fishery has been declining over the past decade. This leaves the natural biomass harvesting and seasonal wetland agriculture as the main sources of livelihood. Over the last five years increasing pressure on the natural wetlands for agriculture has been evident from the rate of encroachment along the shores of the lake and the river floodplains (J.Kipkemboi, pers. observation).

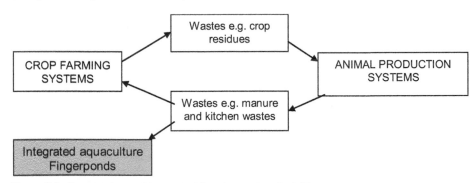

Figure 8.1. Synergy in rural integrated farming system in Africa

The integrated farming system is characterized by diverse farming activities. The advantage of diverse farming system components is the decreased risk associated with a single enterprise. In addition, the different sub-sectors interrelate in a symbiotic and synergetic manner enhancing the overall system productivity. The integration of Fingerponds increases the opportunities for synergy and higher production. Besides fish production, they have additional benefits, such as retention of floodwater in the ponds, which may be used for irrigation, watering of livestock and even domestic water supply. Fingerponds should not be seen as an alternative to existing seasonal wetland use for agriculture by the local communities but rather as diversification of the entire farming system.

Biophysical suitability of Fingerponds

The biophysical suitability for the Lake Victoria littoral and floodplain wetlands for Fingerponds was evaluated in Chapters 2 and 3. The direct use values of the wetlands revealed diversity, ranging from harvesting of natural biomass, seasonal multi-crop production, and other wetland products either for direct consumption by the households or for trade. Among the common wetland uses are the seasonal wetland margin agriculture and flood pool fisheries. This study demonstrated that the seasonal flood pool fishery can be enhanced by designing systems (pond or pools) that can be managed easily to extend fish supply at the household level into the dry season. The ponds and the adjacent gardens are synergistic in terms of nutrient flows.

Fingerponds are highly dependent on natural processes/resources. The wetland soils around Lake Victoria are generally suitable for integrated aquaculture. However, there may be patches with saline soils that affect the general performance of these systems, particularly the cultivation of vegetables.

Another key process driving their functioning is the hydrological cycle. This determines the physico-chemical process and its biota (Figure 8.2). The wetland seasonal hydrological pattern regulates the fish stocking, and the duration of the functional period (Chapter 3). The occurrence of flood events is highly variable between years and depends on the location of the ponds with respect to the source of the flood. Hence, a pre-condition is access to a wetland with regular annual flood cycles.

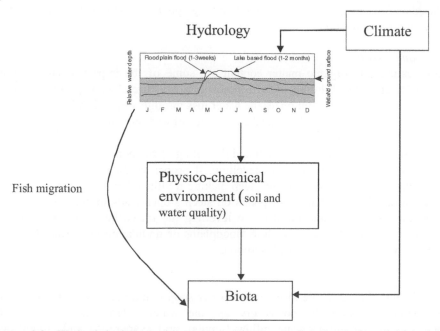

Figure 8.2: Wetland hydrology interaction with the physico-chemical and biological characteristics (Modified from Mitsch and Gosselink, 2000)

In the littoral wetlands and floodplains around Lake Victoria, Kenya the annual flood usually occurs in May. The flood marks the beginning of the Fingerponds' season (Figure 8.3). The culture of fish in the ponds after flood recession may last until around November and December. In January to the beginning of March the ponds may dry up. This allows removal of the sludge accumulated during the previous season. The accompanying gardens can continue as long as there is adequate moisture for crops and the wetland is not flooded.

Figure 8.3: Seasonal calendar of Fingerponds at the Lake Victoria wetlands in Kenya

Additional demand for inputs such as manure other than for Fingerponds is not high and therefore is readily available in most rural areas. Un-used manure from livestock enclosures provides a reliable nutrient source for the system productivity.

Criteria for site selection
Based on 3 years of experimental work on Fingerponds, some observation on site selection for Fingerponds can be made.

Soils and terrain
Soil texture and organic matter content are some of the important characteristic generally considered during site selection for aquaculture. A clayey soil is ideal for pond construction as it minimizes the water loss through seepage. A high percentage of organic matter may be beneficial as a source of carbon, but excess can lead to an oxygen-deficient environment.

Slope affects pond construction and the filling of the ponds by the natural flood. A steep slope requires extensive levelling. If proper landscaping is not done, the embankment gradient may be too steep making it vulnerable to erosion or collapse. In addition, the construction of ponds on steep slopes leads to huge heaps of soil from the excavation and problems in levelling the raised beds. Generally, gently sloping or almost flat areas such as floodplains or undulating lacustrine shorelines are the most suitable sites for Fingerponds.

Flooding cycle
Floods play a crucial role in the initial water supply and fish stocking. After flood recession and until the ponds dry up, the water balance is maintained by precipitation and groundwater inflow (inputs), and evaporation and groundwater outflow (outputs). This may vary from place to place depending on the local conditions. The uncertainty of flood events and the variable length of the season for

fish culture are beyond the control of the farmer and are major challenges. A regular annual flood pulse is a pre-requisite for Fingerponds siting.

Wetland accessibility

Most natural wetlands in Kenya belong to the public or, legally, to the government. They are treated as common property and communities around these ecosystems utilize them for their livelihood. Currently, the dominant activities at the Lake Victoria wetlands are biomass harvesting, livestock grazing and agriculture. Usually access is based on an individual's proximity to the wetland or communal custodianship, if one belongs to the clan "owning" the land adjacent to the wetland. Access rights are amorphous, yet this is critical for the development of the technology, as a considerable investment is needed for construction.

Main benefits of Fingerponds

The potential of Fingerponds in enhancing wetland fishery

African floodplains and littoral wetlands are important for the inland fishery (Welcomme, 1975) but yields are often low and need to be enhanced to meet the growing demands. Fingerponds contribute to the better exploitation of these ecosystems. Their fish yields range from 0.5 to 1 tonne per hectare and can be achieved in a season of about 5 to 6 months. This is equivalent to 10-20 kg per season from a 200 m^2 pond. Machena and Mohel (2001) gave a production estimate of 15 kg per year from 300 m^2 as a typical production level of rural aquaculture systems relying on on-farm inputs. Fingerponds are similar to capture-based aquaculture (CBA), whereby "wild seeds" are obtained and cultured in captivity to marketable size (FAO, 2004). In terms of production, they are intermediate between the capture wetland fishery and conventional, intensive aquaculture (Figure 8.5).

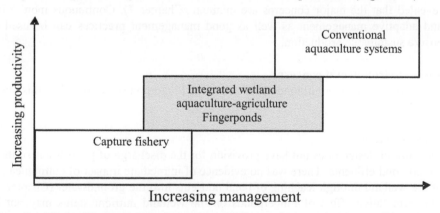

Figure 8.4. Placing Fingerponds in the context of fishery and conventional aquaculture

Contribution to household food security and livelihoods

Aquaculture contributes to food security and poverty alleviation (Bardach, 1985; Tacon, 2001). Fingerpond products consist of vegetables from the raised bed gardens and fish from the ponds. Monitoring of selected households revealed that their per capita annual fish consumption was 5.03 kg with about 93% sourced from the market and 0.33 kg from the seasonal wetland fishery. Based on the ownership of 768 m^2 ponds by 12 households at the Kusa experimental site, the average fish yields from the ponds was 1 tonne per hectare per season, thus they supplied an extra 1.0 kg per capita fish per year. Assuming that individual households will own one Fingerpond of at least 200 m^2 and applying the average yields obtained for a household of seven persons, the potential per capita fish supply results in an additional 3.0 kg per capita per year. This is equivalent to 18% of the world's per capita supply average and 38% for Africa (FAO, 2004). Under good management and with effective harvesting, even higher per capita supplies can be achieved. Considering the current levels of malnutrition of many rural children in sub-Saharan Africa, Fingerponds can contribute sustainably to the fight against food poverty. They improve the rural household livelihoods through their contribution to the social, physical, natural and human assets. In a region where the majority of the rural households are poor, the additional diversity in the resource base is essential for survival.

Potential for wetland wise use

The 3rd meeting of the Conference of the Contracting Parties in Regina, Canada in June 1987 defined wetland wise use as " the sustainable utilization for the benefit of humankind in a way compatible with the maintenance of the natural properties of the ecosystem" (Ramsar Convention Secretariat, 2004). Wise use requires that the exploitation of the use values should not compromise the non-use values. As a result, attention on the environmental impacts are required. Fingerponds enhance wetland fish production. The integration with vegetable gardens adds value to these production systems. With increased intensification, a higher productivity can be realized. This can be used as an alternative to extensive conversion and destruction of the wetland for crop production. Like any human activity, there are likely to be some impacts on the environment. However our environmental impact assessment revealed that the major concerns are minimal (Chapter 7). Continuous monitoring and adaptive management as well as good management practices can be used to ensure sustainable production.

Some environmental concerns

Some of the major environmental concerns of Fingerponds evaluated in this study include:

Eutrophication

The current design does not have provision for the discharge of pond water so there are no pond effluents. There was no evidence of immediate impact of eutrophication of the wetland through leaching of pond nutrients into the groundwater (Chapter 7). The cumulative effect of Fingerponds on the wetland nutrient status may not be manifested over a short period, therefore constant monitoring is required.

Biodiversity

The impact of Fingerponds on the wetland biodiversity was low and was restricted to invasion by seasonal weeds of arable farming on the vegetable gardens. As the system relies on seed fish from the natural environment, there is no threat from escape of un-desired species. During the intra-seasonal harvesting, undesired fish species for culture in the ponds such as haplochromines, *Ctenopoma* sp., and

Aplocheilichthys sp. can be returned to the lake for re-stocking and maintenance of biodiversity. This represents an additional benefit of the technology and re-enforce the natural value of wetlands.

Contamination of the ponds by manure application

According to information obtained at the study sites, the use of antibiotics on livestock is minimal as many people keep traditional disease-resistant cattle (E. Owuor, Ministry of Livestock and Fisheries (GOK), pers.com). This reduces the worries of manure contamination and accumulation of undesired compounds in the ponds.

Recommendations for Fingerponds management

Good Management Practices (GMPs)

Like in many human activities that impact on the environment, a precautionary approach is essential for the reduction of environmental costs. Mitigation of environmental impacts of aquaculture can be attained through Best Management Practices (BMPs) (D'Arcy and Frost, 2001). With respect to sustainable management of Fingerponds systems, the following good management practices are recommended:

- Antibiotics and genetically-engineered feeds, e.g. soya additives should not be used.
- Use of external inputs, especially commercial feeds, should be avoided. Stimulating the natural fish food should be encouraged with the use of farm wastes.
- Introduction of alien species (fish and aquatic macrophytes) is prohibited.
- The use of chemicals for pest or disease control within and around the ponds is to be discouraged.
- Fingerponds should be kept as close to nature as possible. Activities that interfere with the key wetland properties such as hydrology must be avoided.
- Manure use should be restricted to just the amount needed to maintain adequate algal biomass and should be adaptive to avoid deterioration of water quality

Adaptive management approach

Although there is still a need to learn how Fingerponds will work at the farmer-operated level, an adaptive management approach is a good option at the implementation stage. The lessons learned can be used for future improvement. For instance, if a net accumulation of nutrients is detected, the fertilization of the ponds can be decreased in subsequent seasons to avoid excessive accretion. Seasonal manure application rate should be adaptive so that as water levels decline towards the end of the culture period, doses are reduced. The application regime should be balanced with the labour effort available and a daily, weekly or fortnightly application frequency may be adopted.

Sustainability
The goal of any wetland farming system activity is to achieve sustainability (Ramsar Convention Secretariat, 2004; Dixon and Wood, 2005). Kautsky et al. (1997) argue that if aquaculture development is to be ecologically sustainable, efforts must be directed towards methods that make use of the natural environment without severely or irreversible degrading it. In Fingerponds, external input costs such as purchase of fingerlings and supplementary feeds are avoided to keep them economically viable in a rural household setting. The natural fish food is stimulated through addition of just enough nutrients from manure (recycled wastes).

Policy issues and challenges

Land tenure, land use and socio-cultural issues
Land ownership and use-rights for the Kenyan Lake Victoria wetlands is complicated. During this study, it was observed that there was no policy on wetland utilization for livelihoods. Such a situation creates uncertainties and in some cases room for irresponsible behaviour. Wetlands are a common property resource, thus become vulnerable for potential misuse. The culture of "we" which used to be the cradle of many African rural communities and contributed to the protection of common resources is fading. The attitude of "I" seems to have taken over and the communal sense of ownership is quickly fading away due to greed. Selfishness and voracity dominate today's society such that while it is thought that poverty is the major threat to sustainable use of natural resources, a few rich greedy individuals who take advantage of common resources and maximizing the exploitation for their own selfish gains can have major impacts on the environment. The breakdown of social capital is one of the contributors to food insecurity in Africa (Misselhorn, 2005). Again, some of the traditional values which contributed to the protection of the natural environment e.g. the presence of sacred sites, are being discarded as they have no place in the contemporary society. There is need for a clear policy on wetland use and tenure that will ensure equity while accommodating the existing traditional, livelihoods and conservation values. This will not only promote equity among the users but also sustainable use of wetlands.

The initial investment in construction of Fingerponds
Fingerponds is a promising technology for the majority of the poor rural communities living around seasonally flooded wetlands in sub-Saharan Africa. However, the initial investment required for this technology, particularly pond digging may be a limitation to adoption by the poorest households. This requires institutional input particularly from government and Non-Governmental Organizations (NGOs) and from other local community support groups who will play a critical role in lifting this constraint. The construction can be done jointly while ownership can be individual or shared by a group of households.

Gaps in knowledge, inadequate and poor implementation of policies
The ecological and socio-economic database of many wetlands in Africa is still inadequate. As a result, policy issues regarding wetlands are lagging behind, and wetland wise use strategies are still remote. Many countries do not have policies on wetlands or, if present, are fragmented between various departments and are non-

effective. This was recognized by the 7th and 9th conferences of the parties of the Ramsar Convention in Costa Rica in 1999 and Uganda in 2005, respectively (Ramsar Convention Secretariat, 1999, 2005). There is also need for multi-stakeholder partnerships consisting of wetland communities and community-based organizations, government and non-government agencies and research institutions to support development and adoption of the technology.

Coping with uncertainties

As flooding is a natural event, the dependence on it for the initial filling of the ponds and natural fish stocking creates uncertainty. This is a major challenge for the functioning of Fingerpond systems. Deepening the ponds enhances storage capacity and consequently lengthens their functional period. In addition, the layout and spatial location of the ponds in the wetland can be improved to enhance the chances of regular water supply and fish stocking by natural floods. There is need for further study on pond hydrology especially with respect to seasonal variability.

Current trends in wetland resource use and scaling up of Fingerponds

Over the past years terrestrial production has become highly uncertain due to erratic rainfall patterns, thus many wetlands, e.g. Kusa within the Nyando wetland, are threatened by extensive encroachment for crop production and natural biomass harvesting mainly for mat-making (Figure 8.5).

Figure 8.5: Evidence of the pressure of human activities on natural wetland resources due to high demand for (a) cultivated and (b) natural biomass production

A conservative estimate using a wetland margin of 8.5 km and an average population density of 250 persons km^{-2} indicates that uncontrolled wetland agriculture can be a major threat to the wetlands around Lake Victoria. Assuming all households living within a 3 km stretch from the wetland margin cultivate the wetland at the current rate of 0.5 hectares per household, and that 90% of the households will utilize the wetland for seasonal crop cultivation, about 410 hectares of wetland will be converted to crop production. For papyrus biomass harvesting, and using the Kusa wetland as an example, a total of 50 hectares (1.5 %) of the wetland is required. Fingerponds would require about the same area as papyrus biomass harvesting. However, assuming each household has one pond of 10 by 20 m and a garden of similar size, a total stretch required will be about 17 km of wetland margin. In practice Fingerponds should be scattered and it is assumed that not all households will venture into integrated wetland aquaculture. In scaling up, it is important to consider all wetland functions, e.g. livestock grazing, biomass harvesting and natural functions (flood control, buffering, habitat provision for wildlife and carbon sequestration, etc.).

Whereas other wetland land use for livelihoods may have potentially limited effects on the wetland, seasonal wetland agriculture requirement leads to extensive change in land use if it becomes the mainstay of the household livelihood (Table 8.1). There is a need for a compromise between the multipurpose utilization for livelihood and other wetland functional values.

Table 8.1: Some land use options, benefits and the effect on wetland ecological character using the example of Kusa (part of the Nyando) wetland at the Lake Victoria, in Kenya

Major land use options	Approximate area required (ha)[a] (fraction of the wetland)	Main direct benefit	Total estimated economic benefit (KES) [b]	Potential effect on the wetland
Papyrus harvesting	50 (1.5 %)	Biomass	5,928,524	Variable but less extensive
Seasonal agriculture	410 (14%)	Food	29,351,556	Extensive or wide spread
Fingerponds	50 (1.5%)	Food	3,119,597	Limited (intensive)
Grazing	n.d	Fodder	n.d	Limited to wetland margin

[a]Estimate based on the assumption that 90 % of households within 3 km away and about 8.5 km stretch along the wetland margin will use the wetland as the main source of livelihood

[b] Estimates based on market pricing

Further research

This study presents a multi-disciplinary approach to the analysis of smallholder wetland aquaculture from a systems perspective. Four years is inadequate especially when dealing with natural ecosystems considering that the effective data collection is only 2 to 3 years. Some of the research and policy issues that came out during study and deserving attention in future are:

1. There is need for a comprehensive ecological-economic analysis of natural wetlands (there is little information about the region's wetlands at the moment) and the establishment of sustainability indices that can be used as guidelines for wetland wise use. This approach should recognize the multiple use of these ecosystems.
2. Quantification of changes in wetland land-use to determine the rate of loss is required. This requires remote sensing (Geographical Information Systems maps), quantitative sampling and participatory research.
3. The environmental impacts of wetland livelihood activities require appraisal, continuous monitoring and periodic review.
4. Nutrient flows can be used to quantify the sustainability of farming systems. One of the limitations of the present study is the lack of precise

data on quantities of nutrient gains and losses at the sub-system and farming system level.

5. There is need for further research on up-scaling Fingerpond technology and how it fits into long-term sustainable wetland management. Classification for wetlands potential for production functions including integrated aquaculture (Fingerponds) is a necessary tool for wetland resource management (McCartney et al., 2005).

6. The adequacy and effectiveness of national wetland policies in the management and conservation of wetlands also deserves consideration in future research.

Conclusion

Wetland resources dominate the livelihood assets of many riparian households. However, there are increasing pressures on wetland resources due to increased poverty and unreliability of terrestrial production, particularly through climatic vagaries. This has led to non-sustainable, destructive encroachment and harvesting from wetlands around Lake Victoria. To prevent this, the overall productivity of riparian farming systems adjacent to wetlands has to be increased to add value to the existing natural biomass harvesting. Sustainable wetland production systems have a potential contribution towards the Millennium Development Goals (MDGs) particularly through food supply. This is particularly important for sub-Saharan Africa; a region beset by poverty, hunger and malnutrition. Fingerpond systems are designed to address this challenge.

Fingerponds are an agriculture-aquaculture semi-intensive/extensive farming system for sustainable food production at the edge of wetlands that respect nature conservation and the wise use of wetlands. Constructed at the edge of wetlands the ponds successfully become self-stocked by native fish from surrounding water-bodies during seasonal flooding and then retain the fish crop for subsequent culture. Its concept rekindles the wetland fishery, which has declined over the years and enhances agricultural production. The dependence on natural wetland processes (flooding for water supply and fish stocking) makes Fingerpond systems economically attractive but presents a degree of uncertainty due to spatio-temporal variability (e.g. in periods of serious drought floodwaters may not reach the ponds). Careful site selection is the key to success. The success of the technology also lies in its integration into the existing household activities so as to reduce the inherent risk in farming especially associated with climate change.

Fingerponds contribute to household food security through their high protein supply per hectare and increased overall yields. With the addition of dung and/or green manure to the pond fish cultures after flood recession a poor homestead of 7 people with one 200m^2 pond can, on average profit each year from an additional food supply of 20 kg of fish and 340 kg of vegetables (*Brassica oleracea* kales locally known as *sukuma wiki* from the accompanying Fingerpond garden). By being incorporated into, and adding diversity to a riparian rural farming system they provide a more balanced diet, especially through increased fish protein production and vegetable vitamins from the horticultural gardens in the lean, dry season.

Natural wetlands in sub-Saharan Africa contribute to nutrient flows through links with the adjacent terrestrial agroecosystems. Fingerponds promote a synergistic link through the assimilation of farm wastes, particularly manure and kitchen wastes.

The outcome is increased diversity in farming system enterprises, nutrient flows and food production. In order to enhance the sustainability of the overall system there is need for intensification. At a small-scale and intermediate production level, intensification will result in increased productivity, and consequently reduced pressure on wetlands caused by extensive and damaging conversion to crop production. The negative impacts of Fingerponds on the environment are minimal and hence can contribute to sustainable food production. This leads to the conclusion that enhancing wetland fish production and seasonal wetland agriculture through Fingerponds systems may contribute to:

- increase in household food supply and nutritional status
- improved household resilience through diversification of livelihood resources
- improved wetland resource and catchment nutrient management
- reduced pressure on lake shore fishery
- wise-use strategies for wetland conservation

The capital needed for pond construction may limit adoption by poor households. Institutional co-operation and support, particularly from the government, NGOs and other local community support groups is required. It is important to recognize the multiple services and products from wetlands and to integrate them into management strategies for sustainable livelihoods. Further development, therefore, requires institutional collaboration through multi-stakeholder partnerships; participatory research on integration into existing farming systems, up-scaling and technology improvement, and translation of research results into wetland policies with clear guidelines for communities and decision-makers.

Sustainable development requires good policies for which quantitative information on ecological and socio-economic dimensions of the ecosystems is needed. Our Fingerponds pilot project captures many of these aspects. The next stage is to expand the scope of integrated wetland production systems through demonstration projects. These will exemplify the integrated approach to wetland resource utilization, for example, through Fingerpond extensive fish farming systems and associated seasonal vegetable gardens. The schemes can be incorporated into tools for decision-making on wetland management and conservation and, at the same time, increase our knowledge-base for the wise use of wetlands.

References

Bardach, J.E., 1985. The role of aquaculture in human nutrition. Geojournal 10(3), 221-232.

Bugenyi, F.W.B., 2001. Tropical freshwater ecotones: their formation, functions and use. Hydrobiologia 458, 33-43.

D 'Arcy, B., Frost, A., 2001. The role of best management practices in alleviation water quality problems associated with diffuse pollution. The Science of the Total Environment 265, 359-367.

Dixon A.B., Wood. A.P., 2005. Wetland cultivation and hydrological management in eastern Africa: matching community and hydrological needs through sustainable wetland use. Natural Resources Forum 27, 117-129.

FAO, 2004. State of World Fisheries and Aquaculture (SOFIA). Food and Agriculture Organization of the United Nations, Fisheries Department, Rome, 153 pp.

Kautsky, N., Berg, H., Folke, C., Larsson, J., Troell, M., 1997. Ecological footprint for assessment of resource use development limitations in shrimp and tilapia aquaculture. Aquaculture Research 28, 753-766.

Machena, C., Moehl, J., 2001. Sub-Saharan African aquaculture: regional summary. In: R.P Subasinghe, P. Buana, M.J. Phillips, C. Hough, S.E. McGladdery and J.R Arthur (eds.). Aquaculture in the Third Millennium. Technical Proceedings of the conference on aquaculture in the Third Millennium, Bangkok, Thailand, 20-25 February, 2000, NACA, Bangkok and FAO, Rome, pp. 341-355.

McCartney, M.P., Musiyandima, M., Houghton-Carr, H.A., 2005. Working wetlands: Classifying wetland potential for agriculture. Research Report 90, International Water Management Institute (IWMI), Colombo, Sri Lanka.

Misselhorn, A.A., 2005. What drives food insecurity in Southern Africa? A meta-analysis of household economy studies. Global Environmental Change 15, 33-43.

Mitsch, W.J., Gosselink, J.G., 2000. Wetlands. Third Edition. John Wiley and Sons, New York, pp. 107-154.

Ramsar Convention Secretariat, 2004. Ramsar handout for wise use of wetlands. 2nd edition, Ramsar Secretariat, Gland Switzerland

Ramsar Convention Secretariat, 1999. Guideline for developing and implementing national wetland policies. 7th meeting of the conference of convention on wetlands (Ramsar, Iran, 1971). Ramsar Convention Secretariat, Gland Switzerland.

Ramsar Convention Secretariat, 2005. An integrated framework for wetland inventory, assessment and monitoring (IF-WIAM), Resolution IX-1 Annex E. 9th meeting of the conference of the parties to the convention on wetlands (Ramsar, Iran, 1971). Ramsar Convention Secretariat. Gland Switzerland.

Tacon, A.G.J., 2001. Increasing the contribution of aquaculture for Food security and poverty alleviation. In: R.P Subasinghe, P. Buana, M.J. Phillips, C. Hough, S.E. McGladdery and J.R Arthur (eds.). Aquaculture in the Third Millennium. Technical Proceedings of the conference on aquaculture in the Third Millennium, Bangkok, Thailand, 20-25 February, 2000, NACA, Bangkok and FAO, Rome, pp. 63-72.

Welcomme, R.L., 1975. The fisheries ecology of African floodplains. CIFA Technical Paper 3, 54 pp.

Summary

This study was stimulated by the need for an integrated approach in wetland wise use. Sustainable management is critical for long-term ecosystem health and people's livelihoods. The potential for smallholder integrated agriculture-aquaculture as one of the possible wetland wise use strategies was explored in two sites on the northern Kenyan shores of Lake Victoria: Kusa and Nyangera.

Most riparian communities living along the shores of Lake Victoria rely on wetland farming or harvesting of natural wetland products for their livelihoods. The potential for the enhanced benefits from these ecosystems through smallholder agriculture-aquaculture wetlands systems integrated into existing farming activities was investigated. Ponds were dug in the wetlands and were used for fish production whilst excavated soil was used to create raised bed gardens for vegetable production. These integrated fish/crop production systems are called 'Fingerponds'. Annual floods stocked the ponds naturally. After flood recession, manure from the adjacent village was used to improve pond productivity. Locally demanded vegetables were grown in the gardens.

The predominantly clayey soils around Lake Victoria are generally suitable for aquaculture. The pilot study revealed that earthen ponds dug in the wetland (Fingerponds) can be adequately stocked during annual floods with local fish species (≥ 3 fish/m^2). The dominant migrant fish into the ponds consisted of three species of tilapia (*Oreochromis niloticus, O. variabilis and O. leucostictus*), *Clarias gariepinus* and *Protopterus aethiopicus*. Manure for pond fertilization is adequately available from the local villages. Fingerponds have the potential of enhancing the existing wetland benefits through fish and vegetable production.

Soil analysis indicated that the soil textural class was clay in both sites and was generally suitable for pond aquaculture. Soil electrical conductivity and sodium levels were significantly higher in Kusa compared to Nyangera. The presence of patches of soils with encrustations of sodium salts at the wetland margin in Kusa affected the overall functioning of these systems and emphasizes the need for careful site selection. In Fingerponds, the water supply is un-regulated and the water balance is maintained by natural losses and gains. At the beginning of the season, flood events are critically important for the initial water supply. During the functional period of the ponds (which lasted for about 6 months after flood recession), precipitation accounted for nearly 90% of the total water gains whilst seepage and evaporation contributed to an average of 30 to 70% of the losses, respectively. Seasonal pond water budgets indicated that the losses outweighed the gains leading to a progressive decline of water depth during the dry season. A prediction of the effect of pond volume and weather conditions on the functional period was carried out using a dynamic simulation model. The results indicated that the culture period can be extended by 2½ months by deepening the ponds to an average depth of 1.5 m. Drier weather may accelerate losses and shorten the culture period by 1-2 months.

The effects of livestock manure applications on nutrient dynamics, water quality and fish yields were studied. There was no observable adverse effect of manuring on pond water quality. Regression analysis indicated that site, pond management (manuring) and the environmental and climatic variables explained a large part of the variation in NH$_4$-N, NO$_3$-N and total nitrogen concentrations with adjusted r^2 of 0.64, 0.70 and 0.65, respectively. The explained variance for o-PO$_4$ and total phosphorus was 58% and 61%, respectively. Manuring increased the total phosphorus concentration in the sediment but only had had marginal effects on total nitrogen. The chlorophyll a concentration was higher in manured ponds, reaching an average of 150 µg l^{-1} compared to an average of 27 µg l^{-1} in un-manured ponds. The net fish yields were highly variable between sites and seasons and ranged from 402 to 1068 kg ha^{-1}, the data showing that manuring was advantageous. The duration of the culture period, site variability and manuring explained 82% of the variation in fish yields. Careful fertilization of the ponds with livestock manure can be used to improve fish yields in Fingerpond systems.

The rural farming systems around Lake Victoria are predominantly subsistence with integrated crop and livestock production. The overall objective of introducing Fingerponds was to increase fish protein supply to the households and diversfy farming activities. The rural farming system was characterized using natural transect mapping alongside identification of bioresource flows between the system components. Nutrient flows were analyzed using Ecopath with Ecosim 5.1 software using nitrogen as the model currency. The model result scenario with and without the wetland demonstrated the importance of the natural wetland in the overall agroecosystem nutrient flows. The farming system is characterized by low nutrient throughput associated with low productivity. Nutrient balance at the Fingerponds sub-system level was highly positive compared to maize production, which is the dominant activity in the terrestrial ecosystem. Diversification of the farming system through integration of Fingerponds increases the nutrient flow pathways and functional diversity. However, Fingerponds had minimal impact on the agroecosystem performance indicators such as biomass to throughput (B/E) and production to biomass (P/B) ratios, which are usually used to gauge ecosystem maturity and hence sustainability potential. This is probably because the overall farming systems productivity is low and Fingerponds is a small component of a larger agroecosystem. Nevertheless, modelling such systems with Ecopath provided a better insight to the agroecosystem nutrient flows.

The contribution of Fingerponds rural household livelihoods was evaluated. The strength of this innovative technology in household of livelihood outcomes lies in enhancement of natural, human and social capital. Since the production level is intermediate, the benefits may not be high in the short-term perspective. Economic analysis showed that the gross margin and net income of Fingerponds is about 752 Euros and 197 Euros per hectare per year, respectively. This is about 11 % increase in the annual gross margin of an average rural household around Lake Victoria. The additional per capita fish supply is 3 kg per season or more from a 192 m^2 pond. The potential fish protein supply of 200 kg/ha is high compared to most existing terrestrial protein production systems. Fingerponds have potential contribution to household food security and livelihood. The results of the sensitivity analysis indicated that the biophysical variations, which may occur from one wetland to another, have implications on the functioning and consequently the economic performance of these systems. This reinforces the need for the integration of these systems into other household activities to protect the household from the potential risk.

Fingerponds technology is promising, especially for rural riparian households living around seasonally flooded wetlands in Africa. There are however, challenges to this technology ranging from uncertainty of some of the key drivers of the functioning of this system (e.g. water and seed fish supply) due to reliance on natural processes to policy issues. The initial investment required for Fingerponds construction may limit some poor households. There is need for support from governments and non-governmental organizations (NGOs). There is also need for a clear policy on wetland land tenure systems and land use to create more responsibility and wise use. Finally, there is also need for research on scaling up of Fingerponds.

Samenvatting

Dit onderzoek kwam voort uit de behoefte aan een geïntegreerde benadering voor het duurzaam gebruik van wetlands. Duurzaam beheer is van vitaal belang voor zowel de gezondheid van het ecosysteem als het levensonderhoud van mensen op langere termijn. De mogelijkheden voor kleinschalige geïntegreerde landbouw-aquacultuursystemen als duurzame beheersstrategie voor wetlands werden onderzocht in twee onderzoekslocaties aan de noordelijke rand van het Victoriameer in Kenia: Kusa en Nyangera.

De meeste mensen die aan de rand van het Victoriameer wonen zijn voor hun levensonderhoud afhankelijk van landbouw in de wetlands of het oogsten van natuurlijke wetlandproducten. De mogelijkheden om de opbrengsten van deze ecosystemen te vegroten door middel van kleinschalige aquacultuursystemen geïntegreerd in de bestaande landbouwactiviteiten werden onderzocht. Vijvers voor visproductie werden gegraven in de wetlands. Met de vrijgekomen grond werden teeltbedden voor de productie van groenten aangelegd. Deze geïntegreerde systemen voor landbouw en visteelt heten 'Fingerponds'. De jaarlijkse overstroming van het meer zorgde voor water en vis in de vijvers. Nadat het water zich had teruggetrokken werd dierlijke mest uit het naburige dorp gebruikt om de productiviteit van de vijvers te verhogen. In de groentetuinen werden lokaal gewilde groenten verbouwd.

De voornamelijk uit klei bestaande grond rond het Victoriameer is over het algemeen geschikt voor visteelt. Een voorstudie liet zien dat aarden vijvers in de wetlands (Fingerponds) door de jaarlijkse overstroming met voldoende vis kunnen worden bezet (≥ 3 vissen/m^2). De meest voorkomende vissoorten in de vijvers waren drie soorten tilapia (*Oreochromis niloticus, O. variabilis en O. leucostictus*), *Clarias gariepinus* en *Protopterus aethiopicus*. Mest voor de bemesting van de vijvers was voldoende beschikbaar in de nabije dorpen. Fingerponds zijn in staat de bestaande opbrengsten van de wetlands te verhogen door middel van vis- en groentenproductie.

Analyse van de bodem toonde aan dat de bodemtextuur klei was in beide onderzoekslocaties. De electrische geleidbaarheid en het natriumgehalte van de bodem waren beide significant hoger in Kusa dan in Nyangera. De aanwezigheid van stukken grond met zoutafzettingen aan de rand van het wetland in Kusa had invloed op het functioneren van deze systemen en benadrukte de noodzaak van een zorgvuldige selectie van de locaties voor Fingerponds. De wateraanvoer van Fingerponds is niet gereguleerd en de waterbalans hangt af van natuurlijke aanvoer en verliezen. Aan het begin van het seizoen is de overstroming van cruciaal belang voor de wateraanvoer. Gedurende de functionele periode van de vijvers (die ongeveer zes maanden duurde nadat de vloed zich had teruggetrokken) was neerslag verantwoordelijk voor bijna 90% van alle wateraanvoer terwijl kwel en verdamping verantwoordelijk waren voor respectievelijk gemiddeld 30 en 70% van de verliezen. Waterbalansen voor het hele seizoen lieten zien dat de verliezen groter waren dan de aanvoer hetgeen leidde tot een gestage daling van de waterdiepte gedurende het droge seizoen. Voorspellingen van de effecten van vijvervolume en weersomstandigheden op de lengte van de functionele periode van de vijvers werden gedaan met behulp van een dynamisch simulatiemodel. De resultaten toonden aan dat de teeltperiode verlengd kan worden met ongeveer twee en een halve maand door de vijvers gemiddeld ongeveer anderhalve meter diep te maken. Droog weer zorgt voor een verkorting van de teeltperiode van één tot twee maanden.

De effecten van toediening van dierlijke mest op de nutriëntenstromen, waterkwaliteit en visopbrengst werden onderzocht. Geen nadelige gevolgen van bemesting op de waterkwaliteit werden geconstateerd. Een regressie-analyse liet zien dat onderzoekslocatie, bemesting en omgevings- en klimaatsfactoren een groot deel van de variatie in concentraties van ammonium- en nitraatstikstof en totaal-stikstof verklaarden (*adjusted* r^2 van respectievelijk 0.64, 0.70 en 0.65). Voor orthofosfaat en totaal-fosforgehalte werden respectievelijk 58 en 61% van de variantie verklaard. Bemesting leidde tot een verhoging van de totaal-fosfor concentratie in het sediment maar had slechts een marginaal effect op de hoeveelheid stikstof. De concentratie van chorophyl a was hoger in vijvers met bemesting en bereike een waarde van 150 μg l^{-1} vergeleken met een gemiddelde van 27 μg l^{-1} in vijvers zonder bemesting. De netto visopbrengst variëerde sterk tussen onderzoekslocaties en seizoenen van 402 tot 1068 kg ha^{-1}. Toediening van mest had een positief effect. De duur van de teeltperiode, onderzoekslocatie en bemesting verklaarden samen 82% van de variatie in visopbrengst. Zorgvuldige bemesting van de vijvers met dierlijke mest kan gebruikt worden om de visopbrengst in Fingerponds te verhogen.

De rurale landbouwsystemen rond het Victoriameer zijn voornamelijk gericht op zelfvoorziening door middel van geïntegreerde gewasteelt en dierlijke productie. De globale doelstelling van de introductie van Fingerponds was de verhoging van eiwitproductie voor de huishoudens en diversificatie van de landbouwactiviteiten. De rurale landbouwbedrijfssystemen werden gekarakteriseerd door een dwarsdoorsnede van het systeem in kaart te brengen en de stoffenkringloop tussen de systeemcomponenten te analyseren. De nutriëntenstromen werden geanalyseerd met behulp van de Ecopath with Ecosim 5.1 software waarbij stikstof als model-eenheid werd gebruikt. Modelscenario's met en zonder het wetland lieten het belang van het natuurlijke wetland voor de nutriëntenstromen van het agroecosysteem zien. Het landbouwsysteem wordt gekenmerkt door een lage omzet van nutriënten en een lage productiviteit. De nutriëntenbalans van het Fingerponds sub-systeem was sterk positief vergeleken met de maïsproductie (de belangrijkste bedrijfsactiviteit in het terrestrische deel van het bedrijfssysteem). Diversificatie van het bedrijfssysteem door middel van integratie van Fingerponds verhoogde het aantal nutriëntenstromen en de functionele diversiteit. Daarentegen hadden Fingerponds een minimale invloed op de agrosysteem-indicatoren zoals de productie/*throughput* (P/E) en de productie/biomassa (P/B) verhoudingen, die doorgaans gebruikt worden om het ontwikkelingsstadium en de duurzaamheid van een ecosysteem te kenschetsen. Dit geringe effect wordt waarschijnlijk veroorzaakt door de lage productiviteit van het hele bedrijfsysteem en het kleine aandeel van de Fingerponds daarin. Het modelleren met behulp van Ecopath verhoogde het inzicht in de nutriëntenstromen van dit agroecosysteem.

De bijdrage van Fingerponds aan het levensonderhoud van de rurale huishoudens werd geanalyseerd. De kracht van deze innovatieve technologie ligt in de verhoging van het natuurlijke, menselijke en sociale kapitaal. Omdat het productieniveau op een gemiddeld niveau ligt zijn de opbrengsten op korte termijn niet erg hoog. Een economische analyse toonde aan dat het bruto saldo en netto bedrijsresultaat van Fingerponds respectievelijk ongeveer 752 en 197 Euro per hectare per jaar bedroegen. Dit is een toename van ongeveer 11% ten opzichte van het jaarlijkse bruto saldo van een gemiddeld ruraal huishouden rond Lake Victoria. Het extra visaanbod per hoofd was minimaal 3 kg per seizoen uit een vijver van 192 m^2. Het potentiële aanbod aan viseiwit van 200 kg per hectare is hoog vergeleken met de bestaande terrestrische productiesystemen voor eiwit. Fingerponds bieden dus mogelijkheden om een bijdrage te leveren aan de voedsel- en bestaanszekerheid van de huishoudens. De resultaten van een gevoeligheidsanalyse lieten zien dan de biofysische variatie tussen verschillende wetlands gevolgen heeft voor het functioneren, en derhalve voor de economische prestaties van deze systemen. Dit benadrukt de noodzaak voor integratie van deze systemen met de andere bedrijfsactiviteiten van de huishoudens om ze te beschermen tegen mogelijk risico.

Fingerponds technologie is veelbelovend, vooral voor rurale huishoudens rond de jaarlijks overstromende Afrikaanse wetlands. Enkele belangrijke uitdagingen rond deze technologie blijven bestaan, variërend van de onzekerheid over sommige factoren die het functioneren van het systeem beïnvloeden (zoals water- en pootvisvoorziening) tot meer beleidsmatige aspecten. De startinvestering die nodig is voor de constructie van Fingerponds kan een hinderpaal zijn voor de armere huishoudens. Ondersteuning van zowel overheidsinstellingen als niet-gouvernementele organisaties is nodig. Daarnaast is er behoefte aan duidelijk beleid op het gebied van landeigendom en landgebruik om meer verantwoordelijk en duurzaam gebruik te stimuleren. Tenslotte is ook meer onderzoek nodig naar het op grotere schaal introduceren van Fingerponds.

Acknowledgements

This study has been stimulated by inspiration of various people who have constantly encouraged me throughout my research. My gratitude goes to my promoter Professor Patrick Denny for his guidance and assistance in English. I wish to thank my daily PhD supervisor and co-promoter Dr. Anne van Dam for his guidance in data analysis and systems modelling.

My sincere thanks also goes to the entire Fingerponds team Prof. Patrick Denny, Dr. Anne van Dam (UNESCO-IHE, The Netherlands), Prof. J.M Mathooko, Dr. Nzula Kitaka, Dr. A.M. Magana (Egerton University, Kenya), Dr. Moses Ikiara (KIPPRA, Kenya), Dr. Roland Bailey (King's College, United Kingdom), Dr. Jan Pokorny, Dr. Richard Faina, Dr. Ivo Prikryl (Enki, Czech Republic), Dr. Frank Kansiime (Makerere University, Uganda) and Prof. Yunus Mgaya (University of Dar es Salaam, Tanzania) with all of whom I had face to face interaction and discussions through e-mails. I wish to also thank my fellow PhD/ Fingerponds Project Research Assistants Ms. Rose C. Kaggwa (Uganda) and Mr. Hieromin A. Lamtane (Tanzania) for their co-operation and encouragement. I am also indebted to the following former staff of UNESCO-IHE; Dr. Han Klein, Dr. Johan Rockström and Dr. Johan van de Koppel for their advice at the beginning of my study. Mr. Manyala from Moi University gave a hand when needed. I wish to also thank the two MSc student Mr. Cyrus Kilonzi and Mrs. Hilda P. Luoga for their team work spirit during this study. My life was also spiced by my PhD colleagues and MSc students at UNESCO-IHE who provided the social and academic ingredients.

Tackling a multidisciplinary topic presented numerous challenges particularly new areas such as socio-economics, modelling and hydrology. I had to consult widely. Thanks to Jan Nonner (Associate professor) and Dr. Chiung Ting Chang of UNESCO-IHE and to all scientists who were always willing to give their time and share their expertise with me. I wish to acknowledge the contribution of Dr. Reg Noble of the International Support Group (ISG) Ontario, Canada, to the Sustainable Livelihoods assessment and Dr. Patrick Moriarty of IRC, the Netherlands for assistance in Bayesian networks.

While in UNESCO-IHE, I interacted with various other people whose names I will not be able to mention all. I would like to express my gratitude to Jolanda Boots, Vandana Sharma, Laurens van Pijkeren and the entire students affairs, support and facility management group. Thanks to Vera Schouten and the staff of the Environmental Resources Department who assisted in different capacities

I wish to thank the local communities at the Kenyan Fingerponds sites, particularly the Komolo women group in Kusa and the pupils and teachers of Nyangera primary school together with local leaders for their help and co-operation. Special thanks to the Fingerponds Project-Kenya Field assistants; Mr. William Owino and Mr. Jared Onyango for their unfailing assistance.

This study benefited from financial support by a number of institutions: the European Union Fingerponds project through Contract no. ICA4-CT-2001-10037, the International Foundation for Science, Stockholm, Sweden and Swedish International Development Cooperation Agency Department for Natural Resources and the Environment (Sida NATUR), STOCKHOLM, Sweden, through grant no. W/3427-1, Netherlands Fellowship Programme (NFP) through UNESCO-IHE and the European Union Marie Curie Individual Fellowship Contract no ICB1-CT-2001-80014. Logistical support provided by Egerton University Zoology Department

166

(currently Biological Sciences) and Research Division and also by Sida-Relma through Kusa Pilot project through Mr. Rolf Winberg. Dr. Rockström was also instrumental in the linkage with Relma–Sida.

I also wish to thank my employer, Egerton University for giving me the opportunity to develop my carreer. Thanks the staff of Egerton University who assisted in one way or another. Special thanks to Mr. Kenneth Kosimbei who doubled as driver and technical assistant.

This acknowledgement would be incomplete without a mention of my wife Rose and my daughters Mercy, Faith and Joy for their patience and understanding while I was away. Also my mother, brothers, sisters and friends for love and encouragement.

CURRICULUM VITAE

Julius Kipkemboi was born on 11th July 1971 in Uasin Gishu, Kenya. He attended his primary and secondary education in Uasin Gishu district. He joined Egerton University in 1990 where he enrolled in a Bachelor of Science degree majoring in Zoology/Botany option and graduated in 1995. In 1996 he was employed in Egerton University under the academic staff development programme. He attended a post graduate certificate training in Limnology Mondsee Austria in 1997. He later pursued an MSc course in Environmental Science and Technology and graduated in 1999 at UNECO-IHE, Delft, The Netherlands.

After the MSc course he taught at Egerton University until October 2001 when he enrolled in a PhD programme at UNESCO-IHE Delft, The Netherlands. His study was co-funded by European Union Fingerponds Project, where he served as the Research Assistant/PhD research fellow, the Netherlands Fellowship Programme (NFP) and the International Foundation for Science (IFS). His PhD was a sandwich construction programme whereby he conducted his field work in Kenya, at Egerton University while consultation and writing of the thesis took place at UNESCO-IHE.

Mr. Kipkemboi is member of various professional organisations; East African Water Association (EAWA), South African Society of Aquatic Scientists (SASAQS), and International Association of Ecology (INTECOL). His present contact address is;

Egerton University
Department of Biological Sciences
P.O Box 536, Njoro, Kenya
E-mail: j_kkipkemboi@yahoo.co.uk

PUBLICATIONS AND OTHER SCIENTIFIC CONTRIBUTIONS:

1. **Kipkemboi, J.**, Kansiime, F., Denny, P., 2002. The response of *Cyperus papyrus* (L.) and *Miscanthidium violaceum* (K. Schum.) Robyns to eutrophication in natural wetlands of Lake Victoria, Uganda. African Journal of Aquatic Sciences 27, 11-20

2. Denny, P., **Kipkemboi, J.**, Kaggwa, R., Lamtane, H., 2006. The potential of Fingerpond systems to increase food production from wetlands in Africa, International Journal of Ecology and Environmental Sciences 32(1), 41-47.

3. **Kipkemboi , J.**, van Dam, A.A., Denny, P., 2006. Biophysical suitability of smallholder integrated aquaculture-agriculture systems (Fingerponds) in East Africa's Lake Victoria freshwater wetlands, International Journal of Ecology and Environmental Sciences 32 (1), 75-83.

4. Mathooko, J.M., Mpawenayo, B., **Kipkemboi, J.K.**, M'erimba, C.M., 2005. Distributional patterns of diatoms and *Limnodrilus* oligochaetes in Kenyan dry streambed following the 1999 drought conditions. Internat. Rev. Hydrobiol. 90 (2), 185-200.

5. Bailey, R., Kaggwa, R.C., **Kipkemboi, J.**, Lamtane, H., 2005. Fingerponds: an agrofish-polyculture experiment in East Africa. Aquaculture News, Institute of Aquaculture, University of Stirling, Scotland, UK, pp.9-10.

6. Pokorny, J., Prikryl, I., Faina, R., Kansiime, F., Kaggwa, R.C **Kipkemboi, J.**, Kitaka, N., Denny, P., Bailey, R., Lamtane, H., Mgaya, Y.D., 2005. Will

168

fish pond management principles from the temperate zone work in tropical Fingerponds. In J. Vyamazal (ed.) Natural and constructed wetlands: Nutrients, Metals and Management. Buckhuys Publishers, Leiden, The Netherlands, pp. 382-399.

7. Van Dam, A.A., Kaggwa, R.C., **Kipkemboi, J.**, 2006. Integrated pond aquaculture in Lake Victoria wetlands. In: M. Halwart and A.A. van Dam (eds.). Integrated irrigation and aquaculture in West Africa: concepts, practices and potential. Food and Agriculture Organization of the United Nations (FAO), Rome, pp.129-133.

8. **Kipkemboi, J.**, Mathooko, J.M., M'erimba, C.M., Mokaya, S. Population dynamics of waterbirds in wastewater lagoons in Njoro, Kenya. Egerton Journal: Science and Technology Series (accepted).

9. **Kipkemboi, J.**, A.A. van Dam, M.M. Ikiara, P. Denny. Integration of smallholder wetland aquaculture-agriculture systems (Fingerponds) into riparian farming systems at the shores of Lake Victoria, Kenya: socio-economics and livelihoods. The Geographical Journal (submitted.)

10. **Kipkemboi, J.**, A.A. van Dam, P. Denny. Smallholder integrated aquaculture (Fingerponds) in the wetlands of Lake Victoria, Kenya: assessing the environmental impacts with the aid of Bayesian networks. African Journal of Aquatic Sciences (submitted).

International conference/workshop presentations during PhD study

Kipkemboi, J. 2003. Enhancing wetland benefits through smallholder aquaculture-agriculture systems in East Africa; Fingerponds. Meeting and symposium of the east African-Austrian Water Association (EAAWA), Mukono, Uganda, 11th - 13th December 2003.

Kipkemboi, J., van Dam, A.A., Denny, P., 2003. Biophysical suitability of smallholder integrated aquaculture-agriculture systems (Fingerponds) in East Africa's Lake Victoria freshwater wetlands. 7th INTECOL wetlands conference, Utrecht, The Netherlands, 25th-30th July 2004.

Kipkemboi, J., van Dam, A.A., Denny, P., 2006. Towards sustainable community-wetland interaction: a pilot study on enhancing contribution to livelihoods through integrated aquaculture production systems (Fingerponds) at the Lake Victoria wetland edge, Kenya. Paper presented at the International Wetlands, Water and Livelihoods Workshop at St. Lucia, KwaZulu Natal, South Africa, 29th January- 3rd February 2006.

Contribution to book chapter

Welcomme, R.L., Brummett, R.E., Denny, P., Hasan, M.R., Kaggwa, R.C., **Kipkemboi, J.**, Mattson, N.S., Sugunan, V.V., Vass, K.K., 2006. Water management and wise use of wetlands: enhancing productivity. In: J.T.A Verhoeven, B. Beltman, R, Bobbink and D.F. Whingham (Eds) Ecological Studies, Volume 190. Wetlands and Natural Resource Management.Springer-Verlag Berlin Heidelberg. pp. 155-180.

UNESCO-IHE STEP

Name	Julius Kipkemboi
Department	Environmental Resources
Period	2001-2006
Supervisors	Prof. Patrick Denny, Dr. Anne van Dam, Prof. Jude M. Mathooko, Dr. Moses M. Ikiara
Daily Advisor	Dr. Anne van Dam

UNESCO-IHE
Institute for Water Education

*STEP ITEM	Year	Study load hours
UNESCO-IHE activities		
PhD seminars (presentation)	2004	40
(participation)	2003,2004,2006	60
Lunch time seminars (4)	2001-2006	160
Other presentations (2)	2005,2006	60
Colloquium and workshops		
Colloquium (4 Project scientific workshops)	2001-2006	160
Project management workshop	2005	30
International exposure (Conferences and workshops)		
7th INTECOL conference, Utrecht, The Netherlands 25th -30th July	2003	40
International wetlands, water and livelihoods workshop , St. Lucia, South Africa, 29th January-2nd February	2006	40
Presentations at international forum		
Oral presentation at the 7th INTECOL, Utrecht, The Netherlands	2003	40
Oral presentation at the Wetlands, Water and Livelihoods Workshop, St. Lucia, South Africa	2006	40
In depth studies (Additional training)		
Ecological modelling	2002-2003	452
Sustainable livelihood assessment	2004	40
Environmental modelling	2006	20
Professional skills (Contribution to UNESCO-IHE education programme)		
UNESCO-IHE MSc supervision	2003,2005	210
UNESCO-IHE education programmes	2003,2006	29
TOTAL		1421

* STEP (Supervision, Training and Education Plan)

Printed and bound by CPI Group (UK) Ltd, Croydon, CR0 4YY

21/10/2024

01777094-0018